EMERGENCE

Also by John H. Holland:

Hidden Order: How Adaptation Builds Complexity

EMERGENCE

From Chaos to Order

John H. Holland

BASIC
BOOKS

A Member of the Perseus Books Group
NEW YORK

Many of the designations used by manufacterers and sellers to distinguish their products are claimed as trademarks. Where those designations appear in this book and Basic Books was aware of a trademark claim, the designations have been printed in initial capital letters.

Library of Congress Cataloging-in-Publication Data
Holland, John H. (John Henry), 1929–
   Emergence : from chaos to order / John H. Holland.
     p. cm.
   Includes index.
   ISBN 0-201-14943-5 (hardback) – ISBN 0-7382-0142-1 (paperback)
    1. Mathematical models.   2. Game theory.   3. Artificial
intelligence.   I. Title.
QA401.H527   1998
003'.85–dc21   99-60010
                                                       CIP

Basic Books is a member of the Perseus Books Group

Cover design by Bruce Bond
Text design by Dede Cummings
Set in 10.5-point Baskerville by Carol Woolverton Studio

First paperback printing, March 1999

Find us on the World Wide Web at
http:// www.basicbooks.com

Contents

List of Illustrations

Preface

ONLY TWO YEARS have elapsed since I completed *Hidden Order*, my first book for a general audience. In the process, I learned that writing in this vein is much more arduous than writing a scientific monograph. The core of a scientific monograph is concise exposition that progresses logically from assumptions to conclusions, with little or no elaboration; the writer employs disciplinary shorthand wherever possible. In contrast, a book written for the general, science-interested reader must make few assumptions, and shortcuts are a bane. Metaphors, simple examples, and other paraphernalia that scientists usually suppress while writing, come to the fore.

I would much rather do research than write about it. It did not occur to me while working on *Hidden Order* that I would be engaged in this difficult process again soon. My sanguine outlook was reinforced because of the "one-shot" nature of most of the Incentives that impelled me to write *Hidden Order*. But something happened along the way to its completion.

In that earlier writing I witnessed a parade of systems, and models, in which complexity emerged from simple elements. Soon memories of discussions of the "unreasonable effectiveness" of mathematics began to weave in and out of that parade. While I am no skeptic about the effectiveness of mathematics, that the effectiveness is "unreasonable" is quite a different matter. The source of this conundrum is a particular view of mathematics, that has developed primarily in the twentieth century. Large numbers of mathematicians now emphasize that mathematics is a distinct study divorced from considerations of the physical world. From that stance, it does seem strange that mathematics offers such concise, effective descriptions of the universe. Still it is a view that ignores the origins of mathematics. These intertwining themes began to intrigue me: what makes it possible to write concise expressions that yield a rich harvest of description and prediction?

When I realized that this question could be discussed without large dollops of disciplinary apparatus, the impulse to write the present book became more than a passing fancy. Explaining the question and the attendant problems to a general audience forces a breadth that is too easily bypassed when one relies on the apparatus of a particular discipline. So there I was, once again engaged in a challenging task that did not deliver the (sometimes) exuberant feelings that accompany primary research. The result is before you. I do not at all regret the effort, but it still surprises me a little that I undertook it.

Emergence is a more personal book than *Hidden Order*, not because it relies less on the ideas of others, but because it spends more time on my own "take" on these ideas. Along the way, I have presented my thoughts about the way science, particularly interdisciplinary science, is conducted. I do not think these ideas are especially iconoclastic—many scientists with an interdisciplinary bent would agree with them. Nevertheless, they are phrased in my own terms. I have also included a bit of autobiography, which inevitably makes the book more personal.

As was the case with *Hidden Order*, the other members of the BACH group (Art Burks, Bob Axelrod, and Michael Cohen, abetted by Carl Simon and Rick Riolo) have vetted most of the ideas here, and conversations with Murray Gell-Mann have kept me alert to broader issues. The Santa Fe Institute continues to be a wondrous venue, offering opportunities and discussions the likes of which I have encountered nowhere else. Science thrives there, integrating a whole range of new insights and theory about complexity into the larger scientific framework.

Anyone who has read Alice Fulton's poetry or essays will see her strong influence on my ideas about the relation between poetry and science. In particular, the sections on poetry depend heavily on conversations with Alice over the years, though she probably has reservations about parts of this presentation.

Dan Dennett warned me early on about the thorny issues of

reductionism, and I have taken those warnings seriously; he is also responsible for lining me up with that agent provocateur, John Brockman.

There are two more acknowledgments I would like to make before I retreat to the realm of gratitude, cloaked in anonymity, to all who have contributed along the way. One acknowledgment is simple and direct; the other is complex, as befits a book on complexity.

First, the simple one: Vivian Wheeler's contributions as copyeditor are exemplars for her craft: she has a wonderful ability to rephrase wordy passages into crisp, attractive prose.

Now, the complex one: I'm sure I would not have written this book without the peace, long horizons, and ever-changing scene Lake Michigan presents to my aerie at Fridhem. It would take a better writer than I to convey that subtle mixture. Maurita Holland is largely responsible for the existence of Fridhem, as she is for many other joys in my life. She has an even more direct responsibility for this book's existence: she has read the manuscript many times in various stages. How she maintains a freshness of view and ability to suggest clarifications I do not know, but it is a major asset in my attempts to convey science to a larger audience. I am a fortunate man to have her as my constant companion!

—John Holland
Fridhem
Gulliver, Michigan
June 1997

CHAPTER 1

Before We Proceed

A WONDROUS vine emerges when Jack plants the seed for his beanstalk, and it unfolds into a world of giants and magic harps. When we were children, Jack's miraculous beanstalk wasn't so far removed from the everyday miracles of fall colors and germinating seeds. Now that we're grown, seeds still fascinate us. Somehow these small capsules enclose specifications that produce structures as complicated and distinctive as a giant redwood, the common day's-eye (daisy), and a beanstalk. They are the very embodiment of *emergence*—much coming from little. Nowadays we know that genes in the seed specify a step-by-step unfolding of biochemical interactions, but only fragments of this complex process are clearly understood. Indeed, it is evident that we will not truly understand genes and chromosomes until we understand the gene-specified interactions that take a seed, or a fertilized egg, to a mature organism. In short, we will not understand life and living organisms until we understand emergence.

We can contemplate emergence in another guise if we turn to a seemingly unrelated arena, that of board games. Agreement on a few rules gives rise to extraordinarily complex games. Chess is defined by fewer than two dozen rules, but humankind continues to find new possibilities in the game after hundreds of years of intensive study. As with the seeds, much comes from little.

In a still different arena, Newton's laws of gravity, or Maxwell's equations describing electromagnetic phenomena, have much in common with the definition of a game. Their equations describe the "rules" of games in which "moves" can be made with the help of the tools of mathematics. These moves take us to new equations

and mathematical statements that are consequences of the defining equations. As in the case of games, we uncover possibilities
quite unsuspected by the authors. Newton could not have guessed
that his equations would reveal the gravity-assisted boost that takes
space probes to the outer planets, and Maxwell, for all his insight,
could not anticipate that his equations would make possible the
exquisite control of electrons that is the sine qua non of electronic
devices. Like Jack's beanstalk, these equations reveal marvels. Indeed, much of our understanding of the physical world emerges
from a small corpus of fundamental equations built on the foundations laid by Newton and Maxwell.

The hallmark of emergence is this sense of much coming from
little. This feature also makes emergence a mysterious, almost
paradoxical, phenomenon smacking of "get rich quick" schemes.
Yet emergence is a ubiquitous feature of the world around us.
Mundane activities such as farming depend on rules of thumb
for emergence—for example, knowing the conditions that influence the germination of seeds. At the same time, human creative
activity, ranging from the construction of metaphors through innovation in business and government to the creation of new
scientific theories, seems to involve a controlled invocation of
emergence.

We are everywhere confronted with emergence in complex
adaptive systems—ant colonies, networks of neurons, the immune
system, the Internet, and the global economy, to name a few—
where the behavior of the whole is much more complex than the
behavior of the parts. There are deep questions about the human
condition that depend on understanding the emergent properties
of such systems: How do living systems emerge from the laws of
physics and chemistry? Can we explain consciousness as an emergent property of certain kinds of physical systems? We will not
know the limitations of scientific answers to questions like these
until we understand the whys and wherefores of emergent phenomena. The central objective of this book is to provide convincing evidence that scientific investigation will greatly increase our
understanding of emergence.

Where Are We Going?

Despite its ubiquity and importance, emergence is an enigmatic, recondite topic, more wondered at than analyzed. What understanding we do have is mostly through a catalog of instances, augmented in some cases by rules of thumb such as "Place the seed in damp soil" or "Get your major pieces in action." Indeed, our present understanding of emergence is often little better than the child's invocation of Jack Frost to explain the wondrous colors of autumn. Such an explanation stirs our imagination but is ultimately unsatisfying. Our instinct is to start looking for a deeper explanation, an explanation that may go as far as the molecular biologist's contemplation of the tangled biomolecular interactions that produce autumn changes. The deeper explanation, once understood, inevitably gives the imagination an exhilarating boost. But just what is it that is being investigated?

It is unlikely that a topic as complicated as emergence will submit meekly to a concise definition, and I have no such definition to offer. I can, however, provide some markers that stake out the territory, along with some requirements for studying the terrain.

First of all, I will restrict study to systems for which we have useful descriptions in terms of rules or laws. Games, systems made up of well-understood components (molecules composed of atoms), and systems defined by scientific theories (Newton's theory of gravity) are prime examples. Emergent phenomena also occur in domains for which we presently have few accepted rules; ethical systems, the evolution of nations, and the spread of ideas come to mind. Most of the ideas developed here have relevance for such systems, but precise application to those sytems will require better conjectures about the laws (if any) that govern their development.

There may be other valid scientific uses for the term "emergence," but this rule-governed domain is rich enough to keep us fully occupied. This book will demonstrate again and again that a small number of rules or laws can generate systems of surprising complexity. Moreover, this complexity is not just the complexity of random patterns. Recognizable features exist, as in a pointillist painting. In addition, the systems are animated—*dynamic*; they change over time. Though the laws are invariant, the things they govern change. The varying patterns of the pieces in a board game, or the trajectories of baseballs, planets, and galaxies under Newton's laws, show the way. The rules or laws *generate* the complexity, and the ever-changing flux of patterns that follows leads to *perpetual novelty* and emergence. Indeed, in most cases we will not understand these complex systems until we understand the emergent phenomena that attend them.

Recognizable features and patterns are pivotal in this study of emergence. I'll not call a phenomenon *emergent* unless it is recognizable and recurring; when this is the case, I'll say the phenomenon is *regular.* That a phenomenon is regular does *not* mean that it is easy to recognize or explain. The task can be difficult even when the laws underpinning the dynamics are known. In chess it took centuries of study to recognize certain patterns of play, such as the control of pawn formations. Once recognized, these patterns greatly enhance the possibility of winning the game. Similarly, it took centuries of study to extract some of the dynamic patterns inherent in Newton's laws, such as the gravitational boosts used in planetary exploration. And still we learn.

Understanding the origin of these regularities, and relating them to one another, offers our best hope of comprehending emergent phenomena in complex systems. The crucial step is to extract the regularities from incidental and irrelevant details. For example, we may use an idealized form of billiards to gain insights into the way colliding molecules in a gas give rise to measurable regularities such as temperature and pressure (more about this in Chapter 9). Or we may use a mathematical description of poker to discern the complexities of political negotiations. This process is called *modeling.*

Although model building is not usually considered critical in the construction of scientific theory, I would claim that it is. Every time a scientist constructs a set of equations to describe the world, such as Newton's or Maxwell's equations, he or she is constructing a model. Each model concentrates on describing a selected aspect of the world, setting aside other aspects as incidental. If the model is well conceived, it makes possible prediction and planning and it reveals new possibilities. Because modeling is so important to this study, the next section provides a prologue about modeling. Then, in Chapter 2, we look more closely at scientific modeling by examining games and maps as historical antecedents. Chapter 3 adds dynamics by looking more carefully at games and the complex models made possible by computers. As the book unfolds, modeling will be a recurrent topic.

The possibilities for emergence are compounded when the elements of the system include some capacity, however elementary, for adaptation or learning. In Chapter 4 we look at a *learning* checkersplaying program, which still eclipses most later work when questions of emergence are foremost. This program *learned* to beat its designer—clearly a case of more coming out than was put in. It did this by making small, experience-based changes in its elements, and it finally extended its abilities to the point of playing at tournament level. The program is fully reducible to the rules (instructions) that define it, so nothing remains hidden; yet the behaviors generated are not easily anticipated from an inspection of those rules.

It is tempting to take the inability to anticipate—surprise—as a critical aspect of emergence. It is true that surprise, occasioned by the antics of a rule-based system, is often a useful psychological guide, directing attention to emergent phenomena. However, I do not look upon surprise as an essential element in staking out the territory. In short, I do not think emergence is an "eye-of-the-beholder" phenomenon that goes away once it is understood.

We can get a better idea of what exists beyond the eye of the beholder if we think of the generators of emergent behavior as *agents*. The classic description of agent-based emergence is Douglas Hofstadter's 1979 metaphor of the ant colony. Despite the limited repertoire of the individual agents—the ants—the colony exhibits a remarkable flexibility in probing and exploiting its surroundings. Somehow the simple laws of the agents generate an emergent behavior far beyond their individual capacities. It is noteworthy that this emergent behavior occurs without direction by a central executive. The simulated neural networks of Chapter 5 supply another example. These networks, constructed by modeling the interconnection of large numbers of simulated neurons, provide an interesting contrast to the checkersplaying program, while still exhibiting clear-cut emergent phenomena.

The combined lessons learned from the checkersplayer and the neural networks lead us to a setting that mimics an idea dating back to classical Greece. The Greeks set forth the idea that all machines can be constructed by combining (copies of) six elementary mechanisms (the lever, the screw, the inclined plane, the wedge, the wheel, and the pulley). Herbert Simon, three decades ago in 1969, refined this notion in a way that has a direct bearing on our objectives: his watchmaker tale illustrates the advantage of constructing a watch by first constructing subassemblies of elementary mechanisms, which are later combined into larger assemblies, which in turn are assembled to yield the watch. We can understand, and manipulate, complex systems more readily within such a setting.

Here I use these insights to gain a setting that lets us look at complexity and emergence in terms of mechanisms and procedures for *combining* them (Chapter 7). To do so, we have to extend the idea of mechanism beyond the overtly mechanical. We thereby come closer to the physicist's notion of elementary particles as mechanisms for mediating interactions, as when a photon causes an electron to jump from its orbit around an atom. Mechanisms so defined provide a precise way of describing the elements (agents), rules, and interactions that define complex systems. The setting that results gives us a common way of describing the diverse rule-governed systems that exhibit emergence.

The first benefit of this setting is that we can compare quite different systems and models that exhibit emergence. Therein lies our hope of finding similarities and common rules or laws. With diligence and good fortune, we should be able to extract some of the "laws of emergence." Chapter 8 initiates this process by describing the checkersplayer, central nervous system (CNS) models, and Copycat (an insightful computer-based model of the analogy-making process) within this general setting. The setting makes it obvious that these systems have mechanisms in common, though they are quite different in detail. In particular, we see that mechanisms for recombination of elementary "building blocks" (recall Simon's subassemblies) play a critical role in all three systems. Furthermore, we find that (a) the component mechanisms interact without central control, and (b) the possibilities for emergence increase rapidly as the flexibility of the interactions increases.

These insights focus attention on agent-based models, where mobile "mechanisms" (agents) interact with and adapt to each other. Chapter 9 modifies the setting of Chapter 7 to allow the mechanisms themselves to modify the pattern of interactions, through adaptation to each other. This extended setting encompasses a new set of examples of emergence, from the miniature universes called cellular automata to the billiard-ball models mentioned earlier. By analyzing these examples in the extended setting, we gain a deeper understanding of the critical role of subassemblies in fostering emergence.

The new examples also demonstrate that emergence usually involves patterns of interaction that persist despite a continual turnover in the constituents of the patterns. A simple example is the standing wave in front of a rock in a white-water river. The water molecules making up the wave change instant by instant, but the wave persists as long as the rock is there and the water flows. Ant colonies, cities, and the human body (which turns over *all* of its constituent atoms in less than two years) offer more complex examples. These emergent macropatterns that depend on shifting micropatterns make emergence fascinating, and difficult to study.

Persistent patterns at one level of observation can become building blocks for persistent patterns at still more complex levels. The subassemblies of the watch in Simon's example illustrate the point in a static framework: the elementary mechanisms known to the Greeks—a lever, a wheel, and so on—are the building blocks for the mainspring subassembly. That subassembly is combined with other similarly formed subassemblies, such as the gearing of the watch hands, to form the complex system known as a watch.

At each level of observation the persistent combinations of the previous level constrain what emerges at the next level. This kind of interlocking hierarchy is one of the central features of the scientific endeavor (see Table 1.1). It will lead us into, and out of, the thorny thicket known as *reduction*—roughly, the idea that we can reduce explanations to the interactions of simple parts. Because we are dealing with emergence in rule-governed systems, reduction has much to do with our exploration. Reduction has been repeatedly examined in philosophy, and sometimes in the other humanities, but its connection to rule-governed emergence has not usually been a facet of these examinations (with some important exceptions—see for example Daniel Dennett's *Darwin's Dangerous Idea*). Chapter 10 examines the creative aspect of reduction as it applies to emergence. Here we see that emergence in rule-governed systems comes close to being the obverse of reduction.

Another point is closely related to the creative side of reduc-

Table 1.1

A typical hierarchy of interlocking scientific descriptions. Mechanisms at each level are based on mechanisms at the previous level.

SYSTEM (SCIENCE)	TYPICAL MECHANISMS
Nucleus (physics)	Quarks, gluons
Atom (physics)	Protons, neutrons, electrons
Gases and fluids (physics)	
confined (e.g., a boiler)	PVT (pressure/temperature/volume), flows
free (e.g., weather)	Circulation (fronts), turbulence
Molecule (chemistry)	Bonds, active sites, mass action
Organelle (microbiology)	Enzymes, membranes, transport
Cell (biology)	Mitosis, meiosis, genetic operators
Organism (biology)	Morphogenesis
Ecosystem (ecology)	Symbiosis, predation, mimicry

Mechanisms from lower levels provide constraints and suggest what to look for at higher levels. Proposed additions at any level must be consistent with observations at all levels.

tion. The building blocks of a watch have been familiar since the time of the Greeks, but the watch is an innovation that has been with us for less than two centuries. Why was the watch so slow to emerge when the building blocks are so familiar? Here we come upon a central point about model building, innovation, and the study of emergence: building a model, or developing a theoretical construct in science, is *not* a matter of deduction. The standard deductive presentation of theoretical constructs in science hides the earlier, metaphor-driven models that lead to the constructs.

Earlier, I mentioned the billiard table as a model—a metaphor—for the colliding molecules in a gas. Chapter 11 begins with James Clerk Maxwell's use of a mechanical metaphor to increase his understanding of electromagnetic fields. The whole question of model building and the use of metaphor in the study of emergence is shadowed by a larger question: how do scientists discover the laws and mechanisms that are so effective in uncovering the hidden order in our universe? Scientists rarely discuss this aspect of their work, though Maxwell is a glorious exception. Chapter 11 brings out the close relation between the construction of metaphors and the construction of models. Together, Chapters 10 and 11 place the book's earlier discussion of model building and emergence in the larger arena of creation and innovation.

This quick tour of the chapters to come should make it obvious that the terrain of emergent phenomena is convoluted. Nevertheless, certain terms can serve as landmarks. Watch for:

- *mechanisms* (building blocks, generators, agents) and *perpetual novelty* (very large numbers of generated configurations)
- *dynamics* and *regularities* (persistent, recurring structures or patterns in generated configurations)
- *hierarchical organization* (configurations of generators become generators at a higher level of organization)

And *model building*, the subject of the next section, underpins the whole venture.

In the final chapter we look again at the ground we've covered and work out a map of the landmarks and core concepts. We see how the general setting developed in the middle of the book resolves some of the mysteries associated with emergent phenomena in complex systems. We also look at the mysteries that remain, and at the approaches the general setting suggests for resolving those mysteries.

Models

From earliest times, human endeavor has been directed toward discovering ways to channel a chaotic world. In the beginning were rule-bound sacrifices to the gods—we modeled the world in terms of personalities and ways of propitiating those personalities. Later, we discovered mechanisms (gates, pumps, and wheels) and ways of using them to control parts of the world, and we began to model the world with mechanisms instead of personalities. Eventually, we arrived at complex computer-controlled devices and models, and scientific models that employ abstract mechanisms. Despite this pervasive use of models, the art of model building is not a familiar topic, even to many practicing scientists. We return to the topic again and again, treating it with increasing sophistication as the book progresses.

Among living forms on earth, the construction of objects and scripts that serve as models is a uniquely human activity. The models may be small—the early Egyptians produced exquisite miniatures of animals and boats—or they may be large—that huge immobile arrangement of monoliths, Stonehenge, can model the passage of seasons. It is less apparent that models are a pervasive part of day-to-day activity. Driving to or from work is model directed; we have a kind of internal map of the principal landmarks and turning points along the way. We are typically unaware of this map until we have to search for an alternate route because of construction or traffic. In that search we carry out a virtual experiment, rather than actually testing the alternative routes. Herein lies a major value of models: we can anticipate consequences without becoming involved in time-consuming, possibly dangerous, overt actions. Even scale models (model ships, model planes, model railroads, and so on) enable us to make measurements that would be awkward otherwise. We can use the scale of a model ship to determine the distance on the real ship between the top of the mast and the tip of the bowsprit. Model planes grow into the models used in wind tunnels to determine flight characteristics. Indeed, as we'll see later, models are indispensable to careful experiment.

The word "model" is used with connotations that go beyond maps and scale models, and the word has been so used for some time:

> When we meane to build, We first survey the Plot, then draw the Modell.
> (Shakespeare)

> A [model is] a tentative ideational structure used as a testing device.
> (American Heritage Dictionary)

It is this broader usage that plays a key role in the study of emergence. A model need bear no obvious resemblance to the thing being modeled. Newton's equations, as symbols confined to a sheet of paper, look not at all like the orbits of planets around the sun. Yet they model this physical reality in ways that no scale model of the solar system could achieve. Today we go further, programming computers to model real or imagined situations. The examples range from video games to highly detailed flight simulators. We'll soon see how this is done, but for now simply note that models, above all, make anticipation and prediction possible.

For most of us model building starts at an early age. As children we use building blocks to generate concrete realizations of our imagination—castles and space stations. This facility for recombining standard objects to make new items carries over into later occupations. A watchmaker uses familiar mechanisms—gear wheels, springs, pinions, and so on—to generate marvels of timekeeping, and a scientist does the same thing at a more abstract level, generating complex objects, such as molecules, from simpler objects, atoms. By selecting the building blocks and the ways of recombining them, we set up the rules that make rule-governed systems comprehensible. A well-conceived model will exhibit the complexity, and emergent phenomena, of the system being modeled, but with much of the detail sheared away. That will be a central theme of the middle of this book (Chapters 6–9).

In one sense, all of science is based on model construction. The equations of Newton and Maxwell, after all, model aspects of our physical world, enabling us to derive consequences and make predictions. The unanticipated predictions and marvels tied up in these equations provide some of our best examples of emergence. A great deal more comes out than the authors anticipated, even allowing for their superb intuition. To understand emergence, we must understand the way in which models in science and elsewhere allow us to transcend the knowledge that went into their construction.

Models, especially computer-based models, provide accessible instances of emergence, greatly enhancing our chances of understanding the phenomenon. Moreover, the model can be started, stopped, examined, and restarted under new conditions, in ways impossible for most real dynamic systems (think of an ecosystem or an economy). We come full circle when we build computer-based models, providing the computer with a program that is fully capable of surprising its programmer. Our first detailed example of emergence (Chapter 4) will be a checkersplaying program that learned to play better checkers than its designer. We can dissect this program to the point where nothing remains hidden, yet it provides an undeniable case of emergence.

What Stands in the Way?

In the face of the inducements, and possibilities, offered by a deeper understanding of emergence, it seems strange that the available explanations are so limited. There are philosophers, and some scientists, who think that emergence simply *cannot* be explained in scientific terms. Specifically, they hold that the study of emergence cannot be reduced to the study of well-defined mechanisms and their interactions. Scholars of this persuasion are convinced that machines cannot produce self-generated extensions and improvements, where "more comes out than was put in."

This stance is similar to one widely held until the middle of the twentieth century: that machines cannot reproduce themselves. The reasoning was based on the concept that a machine, to reproduce itself, would need a description of itself. But that description would have to include a description of the description, and so on, ad infinitum. "Clearly" this was an impossibility, not so very different from getting more out than you put in. Because living organisms obviously reproduce themselves, this "impossibility" was taken as a major distinction between machines and living organisms. This position on self-reproduction collapsed in the 1950s when John von Neumann, working with an idea provided by Stan Ulam, provided a description of a self-reproducing machine (von Neumann, 1966). As with the scientific explanation of autumn's changes, there was no diminution in wonder. Instead a new realm was opened, offering new wonders and new questions.

I think the barriers to developing a similar mechanical explanation of self-generated enhancement, and emergence, are not those of principle. The difficulty, it seems to me, stems more from the daunting diversity of emergent phenomena. Like consciousness, life, or energy, emergence is ever present, but protean in form. In part, too, the difficulty stems from the similarities between emergent phenomena and serendipitous novelty. The play of light on waves produces an ever-changing scintillation, but there is little of the organization we would expect of emergence in a rule-governed system. The false trails of serendipitous novelty, alongside the widely different examples of emergence, make it difficult to isolate the elements of emergence.

Getting On with It

The program for studying emergence set forth here depends on *reduction*. Complicated systems are described in terms of *interactions*

of simpler systems (again, see Table 1.1 for familiar examples of reduction in science). I emphasize "interactions" because there is
a common misconception about reduction: to understand the
whole, you analyze a process into atomic parts, and then study
these parts in isolation. Such analysis works when the whole can be
treated as the *sum* of its parts, but it does *not* work when the parts
interact in less simple ways. For example, we can analyze a complex sound wave, say an instant from a symphony, into its component frequencies, and then reconstruct the whole by adding these
components together. Some kinds of digital recording depend on
an instant-by-instant ability to recombine component frequencies
into a sustained performance. However, when the parts interact in
less simple ways (as when ants in a colony encounter each other),
knowing the behaviors of the isolated parts leaves us a long way
from understanding the whole (the colony). The simple notion of
reduction—studying the parts in isolation—does not work in such
cases. We have to study the interactions as well as the parts.

Emergence, in the sense used here, occurs only when the activities of the parts do *not* simply sum to give activity of the whole. For
emergence, the whole is indeed more than the sum of its parts. To
see this, let us look again at chess. We *cannot* get a representative
picture of a game in progress by simply adding the values of the
pieces on the board. The pieces interact to support one another
and to control various parts of the board. This interlocking power
structure, when well conceived, can easily overwhelm an opponent
with higher-valued pieces that are poorly arrayed. A valid analysis
of the game's setting must provide a direct way of describing these
interactions. The same holds, a fortiori, for more sophisticated
versions of emergence.

Because emergent phenomena show up in many different scientific disciplines, our exploration is perforce interdisciplinary.
Accordingly, we can accumulate examples of emergence that are
quite different one from another. Because of these differences, we
can use a general setting to compare the examples and throw out
properties that are incidental. This process enhances our chances
of discovering, and controlling, the essential conditions for emer-

gence. The general setting that serves us well in this endeavor is based on the interaction of mechanisms, a choice justified by the detailed examples in the early part of the book.

Mathematical Notation

A final word before proceeding. This book uses mathematical notation here and there, and even some elementary equations; but one can follow all the main ideas without delving into this mathematical arcana. All such notation is enclosed in boxes, like the one that encloses this comment. The book is written so that the reader can skip these boxes without losing the main thrust of the discussion.

That does raise a question. If it is possible to skip the mathematical parts, why include them at all? The answer comes most quickly by looking at a similar, but more familiar domain: music. Anyone, with a bit of effort, can appreciate quite complicated music, but there are musical subtleties that are difficult to convey without notation. The sophisticated compositions of Bach, Beethoven, and Prokofiev depend on the discipline that produces that musical notation. To know musical notation is to enrich one's understanding of the music and of the process of composition. Mathematical notation is for the scientist what musical notation is for the composer. Much is missed if you have no opportunity to look at the notation. By including some of the mathematical notation, I hope to give the reader an opportunity to see some of the subtleties. It takes more effort, but I think the return far exceeds the effort. The notation is simple, and it is explained; the reader is not expected to know any of the structures and theorems that a practicing scientist, much like a composer, takes for granted.

CHAPTER 2

Games and Numbers

Beginnings

I̶T WAS 2:00 A.M. on a midsummer night in 1952. The two of us were building models, and we were being paid to do so.

We were working in a room filled with hulking boxes, each bigger than a large refrigerator. Most were filled with glowing vacuum tubes. One box displayed an array of glowing cathode-ray tubes, much like small television screens, each showing row upon row of dots. Another had a large control panel filled with toggle switches and little orange diode lights. All the boxes were interconnected by large cables, accessible through panels in the false floor. The room held tons of carefully designed circuits, all working together as the prototype of IBM's first fully electronic digital computer. The preliminary manuals called it the *Defense Calculator*—the Korean War was under way. It was far more impressive than anything in a science fiction movie of the time.

For both of us—Arthur Samuel was my companion—the computer opened possibilities for exploring models far beyond the reach of pencil, paper, and adding machine. Scientific models, prior to the advent of programmed computers, were almost always simple because you could neither analyze nor explore a complex model. Emphasis on simplicity was reinforced by the fact that important physical laws could be expressed with few equations (Newton and Maxwell again). Programmed computers opened new possibilities, because they could execute long sequences of instructions at high speeds. Suddenly it was possible to explore models that were orders of magnitude more complex. This blessing also entailed a curse: you could run amok in detail, to the point that you would lose all possibility of uncovering overarching principles. Despite the curse, the opportunities were enthralling.

Samuel and I were adept at controlling this electronic behemoth because we had both been directly involved in its design. Despite its enormous bulk, the Defense Calculator could store only a few thousand numbers; but it was fast. Not by current standards, but fast nonetheless—it could execute sequences of instructions at the rate of a hundred thousand instructions a second. More important, conditional instructions allowed changes in the course of the calculation as results accumulated, without time-consuming human intervention. The programs, and hence our models, could modify their calculations, and even the instructions defining them, as the programs were executed. Self-modifying programs?! It was a heady thought, with visions of self-reproduction, learning, and adaptation. We took the possibilities seriously.

It was also exciting that we could see the implications of our models unfold as we watched. It usually requires months or years of intense study to understand the behavior of traditional models, models defined by sets of equations. A model defined by a computer program, on the other hand, is more like a recipe. Furthermore, the computer is like an automated stove: once the recipe is inserted, the delicacy described emerges. By using programs to define our models, we could rely on the computer to reveal the behavior implicit in the model's definition.

Computer-based models present the modeler with a rigorous challenge. The claims of verbally described models are often established by rhetoric. What appear to be equally strong arguments often back competing claims for the same model—consider claims about global warming or species preservation. The same can sometimes be said of traditional mathematical models, where even the most rigorous mathematical proofs skip "obvious" steps. It is impossible to skip steps in a computer program. The computer executes each and every instruction in the sequence given. A missing or incorrect instruction will send the program careening away from the modeler's intent. In this, a computer-based model is much like the *working* mechanical model the U.S. Patent Department required in an earlier age. No matter how clever and convincing the descriptions, if the working mechanical model didn't produce the results claimed, the patent was denied. Similarly, a computer-based model is both *rigorously described*—presented as a program that can be examined in detail—and it is *executable*.

The next section takes a preliminary look at the modeling Samuel and I were doing, with the object of providing some explicit instances of the relation between models and the study of emergence. This preliminary look motivates a closer examination of three artifacts: board games, numbers, and "building blocks." Board games provide ancient and direct examples of the way in which great complexity can arise from simple specifications; numbers exemplify the shearing away of detail to get at fundamentals; and building blocks offer a direct way of generating complexity, and emergence, from simple specifications. The concepts that board games, numbers, and building blocks exemplify are central themes of this book.

The last section of this chapter returns to computer-based modeling as a way of integrating these themes into a scientific investigation of emergence. In Chapters 4 and 5, we return to the models Samuel and I constructed, to provide working definitions for these basic themes. That, in turn, will prepare the way for the construction of a general setting within which to examine the wide variety of systems exhibiting emergence.

Checkersplaying and Neural Networks

Art Samuel and I reasoned that if we could harness the possibilities inherent in the Defense Calculator, we could write programs that could change—*learn*—as their calculations explored alternatives! In Samuel's words, we could design programs that would enable us to "tell the computer *what* to do, without telling it *how* to do it." Computer-based models are common now (video games are only the most obvious incarnation), but they were in their infancy then. Programs that learn, changing the course of calculation as the model accumulates experience, are rare after almost half a century of endeavor. We still have little theory to guide us and few implementations. One of the goals of this book is to explain the difficulties that accompany this kind of computer modeling, despite the enormous increase in the raw power of our computers since the '50s.

Though we were in the same room using the same machine, Samuel and I had very different models in mind. Samuel wanted to design a model that could learn to play the game of checkers, improving its performance as it played against a succession of opponents (Samuel, 1959). At the time, I felt that Art Samuel's quest for a learning checkersplayer was interesting and challenging, but too ad hoc to get at fundamental principles. How wrong I was! I'll put off the discussion of Samuel's insights for a while, because we need a better understanding of models and model building to appreciate them. Suffice it for now to say that Samuel's efforts contribute directly to a fundamental understanding of the area he named, *Machine Learning*. The depth of his insights, and the lack of subsequent progress, can be summed up by saying that Samuel's results have yet to be surpassed.

For my part, I had been inspired by a lecture by J. C. R. Licklider (then director of the Advanced Research Projects Agency of the Department of Defense) on Hebb's 1949 neuropsychological theory of learned behavior. Hebb's object had been to build a theory of behavior based on the interactions of neurons in the central nervous system. He proposed a mechanism for changing the strength of connections in the network so that successful behaviors would be reinforced. Because I had been reading papers by McCulloch and Pitts (see Kleene, 1951) and Rashevsky (1948) on the logic of neural networks, Licklider's lecture generated a strong "I have to try this" reaction. My boss, Nathaniel Rochester, agreed; so we proceeded to build a pair of models based on the mechanisms Licklider described.

There will be much more to say about neural networks later, but here is a brief précis. The central nervous system (CNS) consists of cells called *neurons*. When a neuron is sufficiently stimulated, it *fires*, producing an electrical pulse that goes out over a long extension of the neuron called an *axon* (see Figure 5.1 in Chapter 5). This axon forms contacts, called *synapses*, with many other neurons. When a pulse arrives at these synapses, it stimulates the neurons contacted. If enough pulses arrive at the synapses on a neuron's surface within a short time interval, that neuron in turn fires. If one traces a sequence of connections in the CNS, that sequence is quite likely to form a loop that returns to the origin of the sequence. That is, the neuron at the start of the sequence originates a chain of pulses that eventually feed back to restimulate the originating neuron (Figure 5.1). This feedback can make loops of connections reverberate—a kind of "ringing"—without further stimulation. As a result, the central nervous system is continually ablaze with activity, multitudes of pulses circulating even during deep sleep and unconsciousness.

Using these basic facts, Nat Rochester simulated a 69-neuron network that kept track of every pulse generated by the neurons; I built a 512-neuron model that used only the firing frequency of the simulated neurons. It took every trick I could think of to cram 512 neurons into the limited storage capacity of the machine, one of the reasons for the late-night vigils. (With no lack of hubris, I gave my model the code name *Conceptor.*) Both models were designed with the neurons interconnected in a tangled array of loops or cycles. The object was to demonstrate Hebb's basic conjecture that as connections are strengthened under repeated exposure to a stimulus, groups of reverberating neurons form. These groups in the CNS, called *cell assemblies,* respond to and come to represent different environmental stimuli. Later they become the building blocks for more sophisticated reactions to the environment. I was able to demonstrate the formation of cell assemblies in my model, but the work fell far short of delineating the possibilities of neural networks with cycles.

Here again we bump into the half-century hiatus: Hebb's theory is still *the* neuropsychological theory, but we don't yet understand the behavior of cyclic neural networks very well, even though they are a sine qua non for the theory. "Feedforward" neural networks, much in the news these days, eschew interior loops, which makes them largely irrelevant to Hebb's theory.

Mysteries and Deeper Mysteries

If it seems that there are some mysteries embedded in these modeling endeavors, there are. In a sense, these mysteries have been part of the subject since humans first began to build models. Broadly conceived to include such things as maps, games, paintings, and even metaphors, models are a quintessential human activity, and they are often mysterious. It is more than coincidence that many early modeling efforts were under the control of a priesthood. Stonehenge, that giant equinox predictor, is the embodiment of power and mystery.

It is important to resolve some of the mystery that surrounds models and their making. Informal descriptions of model building often act like a magician's diversionary passes, further compounding the mystery. Model building can be a subtle process requiring considerable technique, as does any discipline or art form, but it is not far removed from everyday experience. It can be appreciated with about the same amount of effort that goes into appreciating any composition, be it music, painting, poetry, or science. The resolution of the mystery comes in giving careful attention to the steps involved in building the model.

The mysteries appear at several levels, and there are questions characteristic of each level. At the most elementary level: how can a device that only manipulates numbers model checkersplaying and neural networks? More generally: why has the modeling of some kinds of processes and systems, such as full-fledged learning systems, progressed so slowly? Still more generally: what is it about models that helps us better understand our world? And finally: why are models of all kinds so pervasive, even indispensable, in human activity?

It will become evident that answers to the elementary questions can be fairly comprehensive, but some aspects of the more general questions are likely to remain mysterious for quite a while.

The mysteries just discussed are common to a large array of models—maps, architectural diagrams, scale models, games, flight simulators, mathematical models, cartoons, metaphors, analogies, and mental strategies, to name just a few. The very diversity of this array poses a fundamental question that precedes all others: do these different kinds of models have more than a superficial resemblance?

Models are such an automatic feature of day-to-day existence that we rarely stop to think how ubiquitous, various, and important they are. In our day-to-day activities models are as ordinary as vision and movement, but the ordinariness hides enormous complexity. To penetrate this familiar surface, we must discard the idiosyncratic features of particular models to get at the core features common to all models. If we can extract core features, then we can go on to meld them into a general setting that will guide our exploration. Without a general setting, our role remains that of a taxonomist, a butterfly collector, making a long list of models and their features. Collecting is valuable, but a general setting is necessary if we are to explore coherently the ways in which models help us to understand our world.

This general setting is built upon two cornerstones that have long been a part of human culture, board games and numbers.

Board Games and Rules

Board games are a singular human construct, and were a common feature of the early Egyptian dynasties (3000 B.C. and earlier). Board games are typified by pieces arrayed on a partitioned board, with rules that set the legal ways for placing or moving the pieces on that board. The rules *constrain* the possibilities: not all board configurations are legal, and new configurations follow from legal changes to configurations already achieved. It takes only a few rules to make a game as complex as chess or Go. Moreover, although the rules forbid many possibilities, the number of legal configurations remains large, and the ways of moving from one configuration to another are intricate.

Board games are a simple example of the emergence of great complexity from simple rules or laws. Even in traditional 3-by-3 tic-tac-toe, the number of distinct legal configurations exceeds 50,000 and the ways of winning are not immediately obvious. The play of 4-by-4-by-4, three-dimensional tic-tac-toe, offers surprises enough to challenge an adult. Chess and Go have enough emergent properties that they continue to intrigue us and offer new discoveries after centuries of study. And it not just the sheer number of possibilities. There are lines of play and regularities that continue to emerge after years of study, enough so that a master of this century would handily beat a master of the previous century.

It will become apparent that the rules of a board game have much in common with the rules of logic. From there, it is not a great distance to the axiomatic and equation-based models of science. Much of our modern outlook is conditioned on the discoveries that emerge from this way of looking at the world, from atoms and genes to superconductivity and antibiotics. Mathematical models provide an unusually penetrating way of discovering unexpected aspects of our world. That a modeling technique as abstract as mathematics should be so efficacious is a mystery often noted by scientists. It is a mystery we can begin to resolve when we put it in this context of games and rules.

Numbers

At its foundations, mathematics depends on *numbers,* another of those concepts that is at once familiar and mysterious. Numbers may seem to be the very embodiment of concreteness. After all, what could be more concrete than saying, "There are three buses in the parking lot" or "I have two children." But a careful look at numbers starts with abstraction—shearing away detail.

Numbers go about as far as we can go in shearing away detail. When we talk of numbers, nothing is left of shape, or color, or mass, or anything else that identifies an object, except the very fact of its existence. Another way to say the same thing is to state that all collections of objects that have the same number of objects, say three, are to be treated as equivalent when we are talking about number. Three buses, three storks, and three mountains are equivalent "realizations" of the number three.

Shearing away detail is the very essence of model building. Whatever else we require, a model *must* be simpler than the thing modeled. In certain kinds of fiction a model that is identical with the thing modeled provides an interesting device, as with Borges' (1970) map to the same scale as the land being mapped; but it never happens in reality. Even with virtual reality, which may come close to this literary identity one day, the underlying model obeys laws that have a compact description in the computer—a description that *generates* the details of the artificial world.

We can, of course, change the details sheared away. The color "red" treats as equivalent all collections of objects that have that color. Similarly, we throw away masses of detail when we invent concepts such as "trees," "grandmothers," and "airplanes." An individual tree, for instance, has a plethora of detail about leaf shape or placement of branches, and trees of different species can vary in most of their details. Compare an oak to a pine. Still, certain things are held in common by all scenes containing trees, and it is this common part that enables us to build up the "tree" classification. The same holds true for something as specific and unique as "my friend, Alice," where details of dress, hairstyle, and the like are set aside in order to recognize the person. By ignoring selected details we obtain building blocks that appear repeatedly in our experience of the world.

Building Blocks

Any human can, with the greatest of ease, parse an unfamiliar scene into familiar objects—trees, buildings, automobiles, other humans, specific animals, and so on. This quick decomposition of complex visual scenes into familiar building blocks is something that we cannot yet mimic with computers. The task is too complex to be carried out by brute force, despite the computer's tremendous advantage in speed, and we have no plausible computer-based models of human parsing procedures. This lack of an adequate model is almost certainly related to our lack of understanding of the activities of neural nets with cycles, so that mystery expands into a still larger domain.

Whatever the parsing process, it *is* clear that we can use small numbers of building blocks to construct, or reconstruct, complex scenes and configurations. If we consider vision, we can see the importance of the generative character of building blocks. The actual projection of external scenes on the millions of sensory cells in our eye is never twice the same; nevertheless, every scene has some aspects that have appeared before. Over the years we get better and better at discerning and classifying these common elements—the building blocks. Moreover, because we see the building blocks over and over again, we gain facility in determining their essence, learning just what details are relevant. The same considerations apply, at a higher level, when we consider the tremendous range of expression provided by stringing together copies of the few thousand building blocks we call words. It is our ability to discern and use building blocks that makes the perpetual novelty of our world understandable, and even predictable.

The process of discovering building blocks is a never-ending task. Though the number of building blocks in our repertoire may be small relative to the number of configurations in which they appear, we can always acquire more. Part of this technique is simply refinement of classifications, moving from the general to the more specific. A young child may confuse a cow and a horse, calling both "horsy," while an experienced farmer will distinguish different breeds of cow and will know that Betsy, as an individual in the herd, gets restive when she is milked. Even an experienced camper will learn new building blocks by taking up animal tracking (the newly turned leaf or the displaced pebble) or cross-country trekking in the Arctic (the kinds of snow). Occasionally there is a significant addition to the repertoire of building blocks. In most human activities, the discovery of a major new building block causes a "revolution" and opens new realms of possibility. Think of, say, "perspective" in the arts or "gravity" in the sciences.

As time goes on, humans get better and better at knowing what details to discard. We learn what is irrelevant to "handling" or understanding situations, and we refine our building blocks accordingly. We also learn to use rules—sometimes called laws when they are employed in this manner—to project the way in which the blocks will shift and recombine as the future unfolds. That is, we build models that help us anticipate the future. We even rerun the projections with variations and modifications to see what the possibilities are, and to avoid "falling off cliffs." This use of models is particularly obvious in sophisticated board games, but it comes into play in everything from the mundane task of finding an alternate route when roadwork blocks the usual way home, to the generation of sophisticated hypotheses in science.

Here we come closer to an understanding of the modeling process and its pervasiveness. Equivalences obtained by dropping detail, along with rule-supplied constraints on combination of building blocks, will be key elements of our general setting for the model building process. We will soon look at both ideas much more carefully.

Computer-Based Models

Computer-based models nicely integrate the themes exemplified by games, numbers, and building blocks. To implement a model on a computer, we first determine the model's principal components—the model's building blocks. Then we implement these components as sets of instructions in the computer called subroutines. Finally, the subroutines are combined in the computer in a way that determines their interactions, yielding the overall program that defines the model. The result is a computer-based realization of the rules that define the model's behavior.

Computer-based models are at once abstract and concrete.

They are abstractly defined in terms of numbers, relations between numbers, and changes in numbers over time—a feature they share with mathematical models. At the same time, the numbers are actually "written down" in the computer's registers, rather than being represented symbolically. Moreover, the numbers are overtly manipulated by the computer's instructions, much as a grain mill produces flour. We can produce quite concrete records of these manipulations. These records are closely related to the laboratory notebook records of a carefully run experiment. Computer-based models, then, partake of features of both theory and experiment. This combination of the abstract and the concrete offers advantages and also disadvantages.

It is at first surprising that a wide range of concrete objects and processes can be represented by numbers and the manipulation of numbers. Both computer-based models and mathematical models share this rather mysterious ability. How is it, for instance, that we can use numbers to simulate the flight of an airplane over some well-known piece of real estate, say Chicago, with specified weather conditions, say summer thunderstorms. Numerical representations of this kind have become so common that you can run flight simulations on your home computer. With a bit more effort, we get full-fledged industrial flight simulators that wring sweat from experienced pilots when they "fly" in simulated emergency situations. How can this be? It is a mystery we will have to confront squarely, and solve, when we reach that point in our exploration.

CHAPTER 3

Maps, Game Theory, and Computer-Based Modeling

As we search for deeper understanding of the relation between numbers and models, maps are an appropriate starting point. Maps eliminate detail in a straightforward way and, like games, they are among the earliest model-artifacts. Moreover, our longer-range objective, a general setting for emergent processes, is a kind of map, so a better understanding of maps will help define that objective.

Think first of a simple map, such as a road map (see Figure 3.1). If it is fairly complete, as is true of most state road maps, then the cities, towns, and villages are represented by dots or squares of varying sizes, and the roads connecting these population centers are represented by lines of various colors representing road quality. Some lakes and rivers may be indicated, but in general the map concentrates on population centers and roads. Two kinds of relations are preserved:

1. There is a one-to-one relation between the population centers and the dots on the map. Each city, town, and village is represented by a dot.
2. The dots are arranged on the map in the same configuration as the population centers in the actual geography of the state. That is, larger cities that are close together in the state are represented by large dots that are close together on the map, a town that is close to the state boundary is represented by a

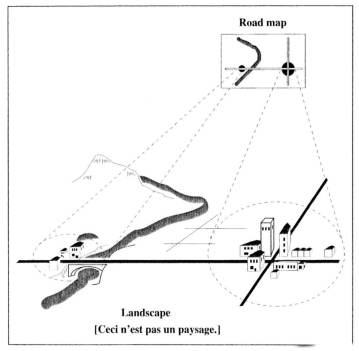

FIGURE 3.1
A road map as a model.

smaller dot close to the edge of the map, and so on. However, all distances have been scaled down, so that cities that are twenty miles apart in reality are separated by two inches on the map. The curves, straightaways, and intersections of the roads are represented on the same scale.

A moment's thought shows that *few* details are retained in this kind of map. We learn little about what we will see at the roadside in driving one of the roads, nor even much about minor zigs and zags in the road (those changes in direction too small to show up at the scale of the map), let alone any details about what the towns look like. What *is* retained is the essential information about getting from one place to another *under normal circumstances*. Road construction or a windstorm can make the route suggested by the map infeasible or impossible.

It is evident that scale plays a major role in the construction of maps. Scale also asserts itself when we extend our view beyond maps to other kinds of models. We at once encounter a whole class of models called *scale models:* scale ships, scale railroads, scale planes. We also expect scale in most statues and representational sculpture, though a monument like Mount Rushmore may be scaled to be *larger* than the original. However, if we look still farther afield, we encounter models in which scaling plays little or no role. Scaling is a special case of a deeper concept, *correspondence*.

We automatically get correspondence when we produce a scaled model, but correspondence is possible without scaling. To construct a model using correspondence, we first select the details or features to be represented, then construct the model so that some part of the model *corresponds* to each selected detail (see Figure 3.1). Think of a cake recipe. It models the steps we actually use to produce a cake. Each step in the recipe (for instance, "add a cup of sugar") corresponds to a complex activity involving a series of physical movements and measurements.

Art Samuel's checkersplayer is a case in point. The correspondence in Samuel's model is between features of the game and parts of his computer program; scale does not enter. For example, corresponding to the "pieces ahead" feature is a set of instructions that actually carries out the counting of pieces. This correspondence between features and computer subroutines will be examined more carefully in the next chapter, after some of the basic ideas are developed here.

Correspondence is best explained with the help of some notation. Let $X = \{x_1, x_2, \ldots, x_n\}$ be a list of details to be modeled, and let $Y = \{y_1, y_2, \ldots, y_n\}$ be the corresponding aspects of the model. Then the correspondence is shown by simply lining up the two sets $\{x_1 \leftrightarrow y_1, x_2 \leftrightarrow y_2, \ldots, x_n \leftrightarrow y_n\}$. In the parlance of mathematics, we have a one-to-one *function, f: X→Y,* mapping details of the object into aspects of the model. The objects on the left (the x's) are called the *arguments* of the function, and the aspects on the right (the *y*'s) are called the *values* of the function. It is interesting that mathematicians use the term *mapping*, as a technical term, when they are being precise in defining functions. The function concept, or mapping, stands at the center of most of mathematics. Because the construction of a model depends on setting up correspondences, the function concept lets us get at the precise heart of model building. It also lets us bring important mathematical tools to bear in our attempts to model systems that exhibit emergent phenomena.

We need not go deeply into the mathematics to see some bonuses from using functions to discuss models. First of all, we can bring numbers into play, with an increase in clarity and precision. It is one thing to discuss "economic health" in rhetorical terms, such as "nervousness in the production sector"; it is quite another to discuss it in terms of the familiar newspaper chart of changes in "gross domestic product" over time (see Figure 3.2). Such a chart matches dollars, as a measure of productivity, against a sequence of dates. This correspondence of numbers to numbers—a function—has enough precision to allow us to determine trends and make forecasts.

We set up a correspondence between the world around us and numbers any time we read an instrument. The numbers on a tire gauge, for instance, correspond to the tire's inflation. Even a calendar is such an instrument, transforming the passage of time into numbers, as did the newspaper chart. It is this transformation that gives instruments and gauges a central role in experimental science. The instruments make it possible to build numerical models of the phenomena being investigated. Because computers are, above all, number manipulators, such transformations are pivotal in constructing computer-based models.

The relation between functions and correspondences also suggests a way of eliminating detail when we construct a model. Features in the checkersplayer are a case in point: *many* boards can share the *same* feature value. For example, there are many boards where the opponent has one more piece than the checkersplayer.

Presentation as a correspondence:

World exports

x (year)	f(x) (trillion $)
1950	0.4
1960	0.7
1970	1.2
1980	2.3
1990	3.5

Graphic presentation:

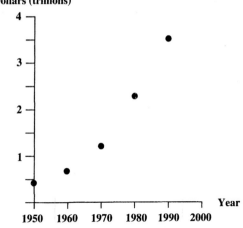

Presentation as an equation:

$$f(x) = 0.4 \times 10^{(t/4)} = \text{world exports at } t$$
$$\text{where } t = (x - 1950)/10$$
$$\text{and } x = \text{year}$$

FIGURE 3.2

Functions and correspondence.

This is a *many-to-one* correspondence. The function that defines the correspondence assigns the same number to many different objects. To use the precise terminology introduced earlier in this section, many *arguments* of the function have the same *value*. In model building, a many-to-one function lets us map objects that differ in detail into a single aspect of the model.

Game Theory

It turns out that there is a close relation between maps and games. This is not so surprising given the maplike character of the boards on which many board games are played, but the relation goes deeper than one might initially suspect. This deeper relation, which will be of great help in formulating a general setting, comes clear in the context of game theory.

Though board games are very old, and games of chance have for centuries played a role in the development of mathematical probability theory, it was not until the first half of the twentieth century that a genuine theory of games came into being (von Neumann and Morgenstern, 1947). Game theory, since its inception, has strongly influenced statistics, information theory, and particularly economics, including recent cross-fertilizations yielding evolutionary game theory (see Maynard-Smith, 1978; Axelrod and Hamilton, 1982). The details of game theory fall to one side of our current exploration, but several concepts from the theory serve us well. For our purposes I concentrate on games that are *not* games of chance, such as checkers, chess, and Go, though much of what I have to say is applicable also to games involving probabilities (chance).

States

The first concept is the *state of the game*. For a board game, this state is simply the arrangement of the pieces on the board at any point in the play. From that point on, the play of the game depends only on that arrangement, not on how it was attained. (There are rare exceptions, such as castling in chess or doubling in backgammon, but these can be handled with the help of an auxiliary "piece," as when a doubling cube is used in backgammon.) In short, the state of the game at any point in the play is a sufficient summary of past history for determination of all future possibilities. In this the state of the game is closely related to the state of a physical system. For instance, we record the state of a container of gas under pressure (a tire or a scuba tank) in terms of its pressure, its temperature, and its volume. If we puncture that container, what happens next is determined by that state. When the state of a system is correctly defined, its future dynamics depends only on its current state.

The *state space* of a board game is simply a collection of all arrangements of the pieces on the board that are allowed under the rules of the game (see Figure 3.3). The qualification "allowed under the rules" is important (see Figure 3.4). In chess, the pieces can be arranged on the board in many ways, but only a small fraction of the arrangements are attainable under the game's rules. For example, the rules of the game require that the piece called a bishop always move to a square of the same color as its starting square (it only moves diagonally on the checkerboard). Moreover, the bishops on a given side start on different colors, so we know immediately that any configuration with these two bishops on the same color is illegal. More carefully: a board game starts with an initial arrangement of pieces specified by the rules; a *move* occurs when the pieces are rearranged under the rules—often the movement of a single piece. Successive moves determine the play of the game. The set of all arrangements (states) that can be attained under the rules is the game's state space (see Figure 3.5).

Tree of Moves

The most important concept from the theory of games, for our purposes, is the *tree of moves* (see Figure 3.5). The *root* of the tree is the game's initial state, the first branches lead to the states that can be attained from the root, the branches on those branches lead to

The state space (set of distinct lawful configurations) for arrangements of black balls in four locations:

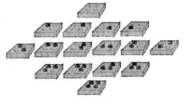

(16 distinct configurations)

Addition of white balls; the new color breaks symmetries in the arrangements of black balls, increasing the number of configurations:

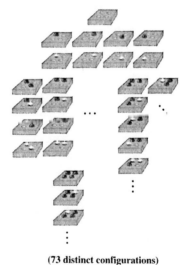

(73 distinct configurations)

FIGURE 3.3
Some simple state spaces.

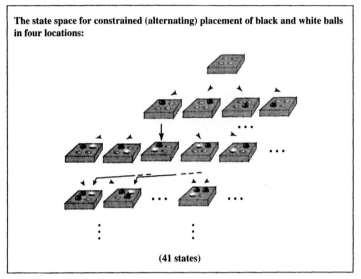

The state space for constrained (alternating) placement of black and white balls in four locations:

(41 states)

FIGURE 3.4
Legal configurations.

the states that can be attained in two moves, and so on to the *leaves* of the tree, which are the ending states. The leaves determine the outcome of the game. It is the succession of choices allowed on the way to a leaf that makes the game interesting.

For real board games, in contrast to some games invented for theoretical purposes, the result is somewhat more convoluted than a tree. In games, unlike trees, different branches may end up at the same state, so there may be fewer states than there are branches. We can move a castle and then a bishop and finish in exactly the same configuration as if we had moved the bishop first and the castle second. In particular, many branches may wind up at the same leaves (end points); in chess, many different lines of play can end with the king checkmated in the corner by a queen and a castle. This additional complication doesn't much affect the present discussion, but I will talk about "ways of playing the game," (instead of about leaves) as a way of indicating this peculiarity.

The rules constrain the game's states (arrangements). The states can be arranged in a tree that shows the order in which the states can be attained.

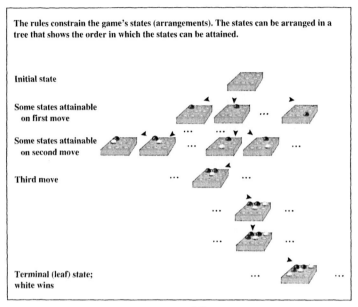

Initial state

Some states attainable on first move

Some states attainable on second move

Third move

Terminal (leaf) state; white wins

FIGURE 3.5

Part of a game tree for tic-tac-toe.

All in all, games are more bushes than trees. The number of leaves (ending configurations) grows *very* rapidly, even when the branching process is simple. Indeed, it is this bushiness that provides the fascination and unpredictability of games. Consider a board game in which there are ten possible moves (branches) from each configuration (state), including the initial configuration. If the game terminates after two moves, there are $10 \times 10 = 10^2 = 100$ distinct ways of playing the game. If the game terminates after ten moves, there are $10^{10} = 10,000,000,000$ ways of playing the game. Termination after fifty moves—a length and number of options roughly equivalent to chess—yields 10^{50} ways of playing the game, a number which substantially exceeds the number of atoms in the whole of our planet Earth.

We begin now to see that a small number of rules can define a game so complicated that we will never exhaust its possibilities. If we had a record of all the games of chess that have been played over the centuries, it is doubtful that any two would have identical move sequences (setting aside those that either terminated early or were deliberate replays of annotated games). It is this perpetual novelty that makes chess and Go classic games that continue to challenge humans after centuries of careful study. By the same token, tic-tac-toe remains a children's game because its possibilities are quickly exhausted, once certain patterns are recognized.

Strategies

In any game that is at all complex, a game plan or strategy is vital for effective play. Roughly, a strategy is a prescription that tells us what to do as the game unfolds; it specifies a sequence of decisions. The game tree provides a way of making this rough idea precise. A sequence of decisions made during the play of a game

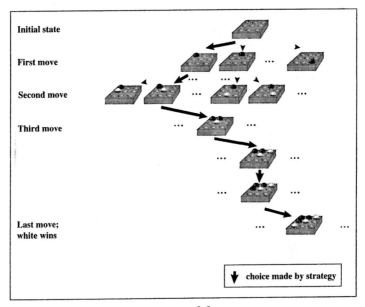

FIGURE 3.6
Opposing strategies determine a path in the game tree for tic-tac-toe.

traces a path in the game tree (see Figure 3.6). So we can define a strategy in terms of the branches it chooses in the game tree. In game theory a *complete strategy* prescribes a branch (move) for *each* state (board arrangement) that can be encountered. In other words, *a complete strategy tells us what to do in any possible situation.* Note that a strategy may be good or bad. It is simply a prescription for what to do; it could be a surefire prescription for losing.

Here is another point where functions are useful. We can use a function to define the correspondence between game states and the moves prescribed by a strategy. The function first assigns a move to the initial state (say, "move the pawn that is fourth from the left ahead one"); it then assigns moves to each of the states that can result from the opponent's response to that move, and so on to the end of the game. For every strategy a function exists that describes that strategy.

More formally, a particular strategy *g* prescribes, for each state *s* in the set *S* of all game states, the decision (move) to be made at that point in the game tree. This is the same as saying that *g* prescribes the follow-on state *s'* in the game tree for each state *s*. For each state *s*, the strategy *g* is constrained to select only branches that lead from *s*. (It may be that a given player will not encounter some states in *S*, but then the prescription can supply a meaningless "dummy move.") In brief, $g(s) = s'$. The strategy, then, is a mapping $g:S \rightarrow S$, where the options for the map are constrained by the set of moves that are legal for each state *s* in *S*.

In a game with two or more players, a *multiperson game,* we can attribute a strategy to each player. Once each player fixes on a strategy, the outcome (leaf of the game tree) is determined (setting aside strategies that use random move choices, say by the roll of dice). To put it another way: the combined strategies select a path through the move tree that leads from the root node to a particular leaf (see Figure 3.6).

When the players have chosen strategies at the outset, it would seem that all the interest and surprises have been removed from the game. All that apparently remains is a kind of mechanical playing-out to reach the predetermined end. But this reasoning omits a factor: the players do not know their opponents' strategies. Each player has decided what to do in each contingency, but each player has no idea what particular contingencies will arise because of the other players' actions. So the individual player *cannot* predict the final outcome, or indeed the outcome of the first few moves, even though that outcome is predetermined. For each player the game will take unexpected twists and turns.

When a game is played repeatedly, the unknown aspects of the other players' strategies may become clearer. Consider a two-person game where the opponent has fixed on a particular strategy. Observing the opponent in repeated plays of the game can tell us what the opponent does at different branch points (choices) in the game tree. We can use this information to build a model of the opponent's strategy. The resulting model will lack many details, because there are just too many possible strategies to uncover a complete description through "trial and error." Nevertheless, if the model is correct in some respects, we can do better with it than without it.

These observations apply to "games" much more general than board games. Consider a game in which one of the opponents is "nature," as when we try to execute a plan (strategy) for enhancing an ecosystem (nature). The outcome may be difficult to predict, even if nature obeys a fixed set of rules (laws). Still, through observation of the effects of choices over time, we can begin to build a model of the ecosystem and its responses. Much scientific endeavor takes this form.

Two major themes emerge at this point:

1. In realistic situations, a strategy *cannot* be defined by listing all the game states with the moves prescribed for each state. Too many states exist in even a modest game to make such a list feasible. This is so even when we take into account the storage capacity and speed of the largest computers; the number of states we calculated earlier for move trees is so large that no foreseeable computer could store them. The recent chess-playing programs provide a direct example. Even with the tremendous storage ca-

pacity and speed devoted to them, they do not attempt an explicit strategy. Rather, the programs are highly selective in the way they search the tree, changing the search as the game unfolds. They test only a minuscule fragment of the whole.

> To say the same thing in mathematical terms, we cannot define the strategy function *explicitly* by listing all (state, move) pairs, $(s, g(s))$, for all states s in S.

Instead of an explicit definition, we define strategies in much the same way we define games, via a set of rules. The rules in the case of strategies are usually rules of thumb. For example, in chess these rules embody principles like "Build a strong pawn formation," "Control the center," "Look for 'fork' attacks," and so on. Such rules pick out game features that occur frequently and are relevant to decisions at various points in the game. In so doing, they group states into clusters, where the states in a given cluster have a feature that suggests similar decisions or moves. In this way we obtain an effective reduction of the enormous size of the game tree and make possible an overall prescription that controls play throughout the game. The extended discussion of Art Samuel's checkersplayer in the next chapter will show how this is done.

From this perspective, we use repeated plays of the game to discover and combine building blocks (rules of thumb, features) in order to construct a feasible strategy. The task is much less daunting than trying to define a strategy explicitly as a list of states and prescribed moves. Even if the strategy has some components that are not easily described in terms of building blocks, it is a valuable starting point for modeling the strategy. This point of view also suggests we assume that the opponent's strategy is constructed from a limited set of building blocks.

If nature is the "opponent"—if we are taking the role of scientists—we do much the same. We attempt to model the rules of the universe, though with less reason to believe that the opponent is restricted to feasible strategies. (Einstein's *cri de coeur,* "Quantum mechanics is very impressive, but I am convinced God does not play dice," is clearly an expression of faith, not an observation.) In science, as in games, part of the justification for making this assumption about building blocks is that it works. Newton's laws, Maxwell's equations, Mendeleev's periodic table, and Mendel's genes all tell us a great deal about the way the world operates.

Note that surprises are in store, even if we should be able to uncover a set of fixed laws that determines all of nature's possibilities. After several centuries we still uncover new possibilities inherent in Newton's equations, and this set of laws most assuredly does *not* encompass all of nature's possibilities.

2. Our simplifying assumption to this point has been that opponents employ fixed strategies, but this simplification sidesteps most of what happens when games are played repeatedly. Opponents learn. A more realistic view is that *all* players are *simultaneously* trying to build models of what the other players are doing. Under this extension, the situation becomes much more complicated. An observer who has an omniscient overview of the game encounters surprises akin to those encountered by individual players. Even if that observer knows the initial strategies and the details of the individual learning procedures, it is next to impossible to predict the course of the game. Emergence and perpetual novelty are ever present in games where the opponents are adapting to each other.

Emergence—A First Look

This view, of opponents adapting to each other's strategies, encourages a more careful look at emergence in rule-governed systems. A computer, once supplied with the rules of a game, and the rules that determine the players' strategies and changes in strategy (setting aside chance moves), can, move by move, determine the course of the game. So the overall system is fully defined. Despite this, an outside observer will be hard put to determine what happens next, even after extended observation. The strategies co-evolve in the computer, each strategy adjusting to its experience with its opponents. This coevolution exhibits the creativity we expect of any evolutionary process; the computer is continually getting into parts of the move tree not previously observed. Species of play rise to dominance and disappear, players mimic each other, and so on. What, if anything, shows the regularity and predictability we expect of emergent patterns?

Though prediction is difficult in these circumstances, it is not a hopeless task. Everything depends on the level of detail we require of the prediction. Meteorology provides a useful simile. Weather patterns are never the same in detail, and even the larger features, such as fronts, cyclones, jet streams, and the like, show a remarkable diversity. Moreover, weather models do not yield exact predictions of such obvious events as the amount of rainfall to expect locally on the morrow. Nevertheless, modern weather prediction is very helpful. It does forecast the likelihood of rain and severe storms, it does give the likely temperature ranges, and the five-day forecasts of average temperature and rainfall are much more accurate than could be attained by simply using past statistics for that time of year.

Chaos theory is often cited as an explanation for the difficulty of predicting weather and other complex phenomena. Roughly, chaos theory shows that small changes in local conditions can cause major changes in global, long-term behavior in a wide range of "well-behaved" systems, such as the weather. In an oft-cited example, the flapping of a butterfly's wings in Argentina can (eventually) cause worldwide changes in the weather. There is a sense in which this is true: if we knew *all* the values for *all* the relevant variables worldwide, à la Laplace (see Singer, 1959), we could predict the weather indefinitely far into the future. With such a model we could determine the long-term weather pattern with and without the flapping of the butterfly's wings. We would see that the two weather patterns would eventually diverge to a point of no correlation.

This explanation ignores important factors in real weather prediction. Because meteorologists do *not* know the values of all the relevant variables, they do not work at a level of detail, or over time spans, in which chaos would be relevant. The predictions work with large masses of atmosphere over short time spans; so butterflies, or jet airplanes, produce negligible effects. Moreover, rather than trying to develop predictions based on remote initial conditions, as with the butterfly effect, meteorologists start anew each day, using the most recent data. These observations continually bring the state of the model into agreement with what has actually occurred. Under this regime chaos theory has little relevance.

The key to effective weather prediction, then, is the discovery and use of the mechanisms (building blocks) that generate weather. This approach originated with the discovery of *fronts* by the Norwegian meteorologist Vilhelm Bjerknes in the early part of the twentieth century. Curiously, Bjerknes lived in Bergen, Norway, where weather prediction is remarkably easy throughout most of the year—it rains! The model that Bjerknes originated has been progressively improved, through the use of more sophisticated mechanisms (and equations) governing fluid flow, the discovery of jet streams, and the recognition that distant large-scale phenomena, such as the Pacific High, can be used to guide long-term predictions. Computer-based models, strongly advocated at the very beginning of the computer era by von Neumann (see Korth, 1965), significantly advanced both the detail and the time span of weather prediction.

Thus, complexity and even chaotic effects need not forestall our study of emergent phenomena. The key to deeper understanding, as with weather prediction, is to determine the level of detail and the relevant mechanisms. At the right level of detail, the model's changing states play the role of the configurations in a game. Using mechanisms as building blocks, we can construct models that exhibit emergent phenomena in much the way that interacting strategies in a game produce patterns of interaction not easily anticipated from inspection of the game's rules. The mechanisms play the role of the game's rules, setting limits to what is possible while providing extensive combinatoric possibilities.

Even when we have the right level of detail and the relevant building blocks, perpetual novelty is still typical. As in games, though the definition is simple, the state space for models of complex systems is very large. And, as in games, the model rarely or never returns to states already visited. This perpetual novelty renders it difficult to make predictions, even when the mechanisms (rules) and the initial state are specified. If the basic mechanisms provide for learning or adaptation, the difficulty increases enormously. Still, by attending to selected details, we can usually extract recurring patterns, like fronts, in the complex unfolding sequence. When these recurring patterns are regularly associated with events of interest, we call them *emergent* properties. We will look much more closely at the prediction of emergence—the "when," "where," and "what"—once we have a general setting in place.

Dynamic Models

In the previous discussion, we have moved from models that have static forms, such as scale models, to models with changing configurations, usually called *dynamic models*. The object in constructing a dynamic model is to find unchanging laws that generate the changing configurations. These laws correspond roughly to the rules of a game. In a game, the rules say how the configurations (states) change as different moves are made; the players affect the course of the game by choosing moves. When we consider the weather system, we usually think of it as autonomous, proceeding without (or despite) our intervention. Still, the laws of change specify the succession of states—the weather configuration eight hours from now, twenty-four hours from now, and so on. If we had effective means of weather control, then the laws of change would specify how those controls affect the unfolding weather sequence.

To build a dynamic model we have to select a level of detail that is useful, and then we have to capture the laws of change at that level of detail. There are potential conflicts. It may be quite difficult to construct a detailed model that is "faithful" to the system being modeled. Weather prediction models provide instructive examples. Predicting that the atmospheric temperature will be less than the boiling point of water may be reassuring, but it is not much of a weather prediction. We want more detail, but then we have to deal with laws of change that involve fronts, jet streams, and the like.

There is, of course, no guarantee that we can find simple laws of change for the level of detail selected. Indeed, the art of model building turns on selecting a level of detail that admits simple laws—a point to which we'll return in later chapters. Setting the level of detail turns mostly on defining the model's *states* (see Figure 3.7). For games, we defined the state of a board game as the configuration of pieces on the board; for Bjerknes' weather model, the current state is the configuration of fronts, jet streams, and the like on the weather map. For dynamic models in general,

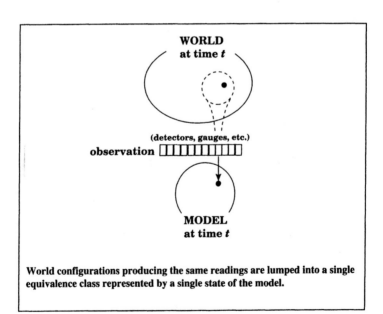

World configurations producing the same readings are lumped into a single equivalence class represented by a single state of the model.

FIGURE 3.7
Observation and state of the model.

the features and details incorporated in the model's states determine the level of detail.

Once we define the model's states, our object is to define the laws of change that work at this level. Laws of change are stated precisely with the help of a *transition function* (see Figure 3.8). The transition function assigns to each state the state that will occur

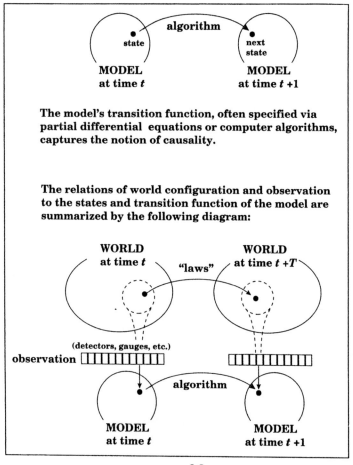

FIGURE 3.8
Transition functions.

next under the laws of change. Where the course of change of the system can be affected from "outside"—where the system can receive inputs from the outside—the transition function provides a different correspondence for each input to each state. That is, different inputs cause different next states, so the transition function provides a correspondence between each possible (state, input) pair and the state that results. The transition function is reminiscent of the function that defines a strategy, where moves are the counterparts of inputs. Newton's equations again provide an example: they define the dynamics of gravity via a transition function that relates mass and acceleration (states of a particle), and force (an input).

If the transition function (law) is "faithful," we can make predictions into the indefinite future. Knowing the current state and input, we can determine the next state. Then, knowing that state and the next input, we can determine the state after that, and so on indefinitely. Therein lies the great advantage of a faithful formal model: by simply iterating the use of the transition function, we can explore future possibilities. The transition function determines the future fully and unambiguously (if the inputs are known). The only uncertainty resides in the appropriateness of the level of detail and in the faithfulness of the transition function. That is, the uncertainty lies in the model's *interpretation*, the mapping between the world and the model.

This capacity for prediction provides the deep connection between modeling and emergence. The (usually simple) specification of a model—the transition function—can yield a limitless array of consequences and predictions. A well-conceived model can, like chess, yield organized complexities that repay decades and centuries of study. Moreover, these complexities may involve possibilities not conceived by the modeler, as when Newton's model is used to guide rockets to Mars and to determine the evolution of galaxies. As with Jack's magic seed, Newton's model opens worlds of wonder that transcend the simplicity of the starting point.

The idea of "faithfulness" takes us from a game, which may only remotely reflect the world, to a model that correctly reflects selected aspects of the world. Interestingly, faithfulness in this sense can be given a simple, precise definition: we say a model is *perfect* if the interpretation satisfies a criterion called *commutativity of the diagram* (see Figure 3.9). Commutativity holds when the order in

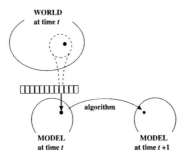

For any world configuration, an observation of the world followed by execution of the model's algorithm

should yield a prediction that matches an observation of the world after a fixed interval of time *T* elapses.

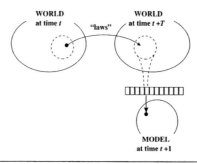

FIGURE 3.9
Commutativity of the diagram—a perfect model.

which we do things is irrelevant to the result. If we take a step right and then a step down, we arrive at the same place as if we take a step down and then a step right. Addition is commutative: $5 + 3 = 3 + 5$. To use this idea for models, we set up a diagram that shows, in its upper half, one time-step of change in the world (say, possible changes in the weather in an eight-hour period). In the lower half of the diagram we show a one-time step change in the model, under its defining law of change (transition function). We can observe the world now (left side of the diagram) or later (right side). In either case that observation determines a state for the model. Commutativity of the diagram holds when, *for every state,* taking the observation (down) and then executing one time-step in the model (right) yields the same result as waiting one time-step in the world (right) and then taking an observation (down). That is, the model's law of change correctly predicts the result of a future observation. Because this definition holds for all states, we can iterate the process to get predictions indefinitely far in the future, as suggested earlier. Of course, we can only sample the states of the world, even if we are only interested in a sharply delimited region (for instance, an experiment). So in practice we can only approximate the "all states" requirement for a perfect model. Nevertheless, the concept of a perfect model supplies a valuable guide for constructing useful models.

Computer-Based Models—A Closer Look

I have mentioned computer-based models several times, and they have a critical role in the construction of dynamic models. They have become ubiquitous in modern science, being used to model everything from the spread of epidemics to fusion in the sun. It will help us to understand dynamic models if we look at computer-based models a little more closely now, though still in preliminary fashion. Earlier I asked how we can use numbers, and coordinated changes in numbers, to simulate the flight of a jet over Chicago in a thunderstorm. I want now to give a partial answer to that question.

The starting point, once again, is the notion of *state*. The natural question is, "What can we possibly mean by the state of a jet airplane flying over Chicago?" The answer is closely connected to the information the pilot uses to fly the jet.

To get at this connection between information and state, let me start with a simpler system: the control panel of the family car. The car's control panel is not in principle much different from that of the jet, it is just much, much simpler. It tells us only the essentials that we need to know when driving, typically the speed of the car, the fuel level, the engine temperature, the battery charge, and the oil pressure. These readings model the state of the car, at a certain level of detail, when it is under way. We could add more readings, such as the air pressure in the tires or the amount of antifreeze in the radiator, to get a more detailed state. This more detailed state would provide the wherewithal for a more sophisticated model; however, decades of experience have shown that the gauges first mentioned are sufficient for operating the car in most situations.

Because the jet is far more complicated, the pilot's compartment is filled with a panoply of displays, gauges, dials, and warning lights that provide information about the conditions that affect the jet's flight. They tell about the plane's speed and position, the amount of fuel in its various fuel tanks, the operating condition of the engines, the position of the landing gear, and hundreds of other bits of data. Indeed, there is enough information for the pilot to fly the plane "blind," using instrument readings alone.

For both the car and the jet, the displays and gauges produce readings that either are numbers or are easily reduced to numbers. A warning light can read either on or off, which can be represented as a 1 or a 0, and even the sophisticated positional display is presented by an array of dots (called pixels), which can be represented as an array of 1's and 0's. In other words, it is easy to reduce the information on the control panels to numbers. These numbers can, as usual, be stored in registers in the computer. Together they define the state of the model, much as the arrangement of pieces defines the state of a board game.

We give the computer a representation of the state of the model by entering these numbers into the storage registers. Then we enter instructions (a program) that cause these numbers to change over time as specified by the transition function. This is the counterpart of defining the rules of the game. The numbers in the registers change in a way that mimics the state changes in the object being modeled. The *universality* of the general-purpose computer assures that any transition function defined by a finite number of rules can be so mimicked.

As in a game, we now confront the notion of choice. The driver or the pilot can choose among alternatives, such as making the car or jet go faster or slower. Phrased in terms of states this means that, once again, from any state we can construct a tree of legal alternatives. In a game, these alternatives were the legal moves allowed by the rules. In the case of the car or the jet, the laws are those imposed by nature and the technology. Executing a sequence of controlling actions is the counterpart of making a sequences of moves in a game. In both cases we choose a path through the tree of possibilities.

When both the numbers and the program have been stored in the computer, we simply start the computer executing its instructions. Think again of a video game or a flight simulator. The instructions, acting on the stored numbers defining the model's state, determine what happens instant by instant. What we see on the screen is a back-translation of the numbers to gauge readings, displays, and so on, that capture the look and feel of the original machine. Controlling actions amount to input to the program at various stages of the calculation. The input is supplied by typing, or by the video game's joystick, or by realistic controls in a full-fledged flight simulator. The result is a dynamic, computer-based model—a major vehicle for the scientific investigation of emergence.

CHAPTER 4

Checkers

Any serious study of emergence must confront learning. Despite the perpetual novelty of the world, we contrive to turn experience into models of that world. We *learn* how to behave, and we anticipate the future, using the models to guide us in activities both common and uncommon. Somehow, through learning, these models emerge from the torrent of sensations that impinge upon us at every moment. Certainly, a deeper understanding of learning will contribute to a deeper understanding of emergence. In attempting to understand the relation between learning and model building, we could scarcely find a better starting point than Art Samuel's mechanization of learning in the checkersplaying program.

It is strange that, until recently, Machine Learning has been a sideshow in Artificial Intelligence (AI)—strange, because most would say that an organism that does not learn is not intelligent. Nevertheless, for most of its history, AI has placed work on learning at the periphery of its activities. Samuel's 1959 work on checkersplaying, and the work on cyclic neural nets (Rochester et al., 1956), both completed almost a half-century ago, still lie close to the cutting edge of research in Machine Learning. Because the two efforts took place near the dawn of the computer age, they have a stark construction, unencumbered by elaborate concern with computer languages, interfaces, and the like. For this reason, they make a meaningful starting point for a close look at the computer-based models that underpin Machine Learning. We will delve into Samuel's checkersplayer in this chapter, and will devote the next chapter to cyclic neural nets. Although these two models are quite different from each other, they have important features in common, features that will help us to arrive at a general setting for emergent processes.

Before we go on, there is an interesting preliminary question: just what is it that Samuel was modeling? He was *not* attempting a detailed model of the thought processes of humans playing checkers. Rather, he was working at the level of strategies. He selected building blocks describing features of the game relevant to good play, and then provided ways of weighting and combining these building blocks to define strategies. Most of all, he was interested in the use of experience to modify and improve these strategies. It is at this level, well removed from neurons or neurophysiological psychology, that the model mimics learning. The principles and rules of thumb uncovered—clarifying the exploration of options (lookahead), subgoals, modeling the actions of other players, and learning in the absence of reinforcement—are at least as important now as at the time Samuel uncovered them. As we'll see in Chapter 6, these principles have a central role in agent-based models of emergence and innovation.

Why It's Difficult to Become a King

By 1955 Samuel had an impressive working model of his basic ideas. Though his 1959 paper on this working model is often referenced, it is difficult to find a careful description of his basic ideas and, indeed, they are not generally part of the armamentarium of the AI community. Nevertheless, Samuel's checkersplayer provides a concrete, and fascinating, illustration of ideas important to an understanding of emergence.

To see why the ideas are important, let's look first at the interlocking problems Samuel confronted in this first attempt at Machine Learning:

1. The programmed checkersplayer had to handle the perpetual novelty of the game.

The number of board configurations that can occur in checkers is very large, far too large to compile a list that programs the response for each possible configuration.

2. The checkersplayer had to learn to make appropriate moves during the play of the game, in the absence of any immediate feedback as to whether the moves chosen were good or bad.

In checkers the sole payoff is at the end of the game, and the information provided then only tells whether the game has been lost or won. During most of the game, it is far from obvious which moves lead to a win (which is what makes the game interesting). Yet a great deal of information is generated as the game is played: information about pieces moved, configurations attained, and so on. The checkersplayer must somehow use this information to develop subgoals (such as having more kings than its opponent) that improve the program's chances of winning.

3. The checkersplayer had to learn about early "stage-setting" moves that make possible later obviously good moves.

Stage-setting is the very essence of winning game play. An example of stage-setting is a move that sets a "trap," making possible a later, obviously beneficial move such as a triple jump. Often an expert commentator can give a description of traps and other "turning points" that set the course of the game. However, human players can usually acquire and use such knowledge without an expert commentator. The checkersplayer must be able to emulate such performance. For example, it must identify configurations and moves that set the stage for later clearly advantageous moves.

4. The checkersplayer had to model its opponents.

The checkersplayer cannot play a winning game of checkers unless it can anticipate its opponent's best moves in various situations.

The difficulty of these obstacles is compounded by Samuel's desire that the program *learn* to overcome the obstacles, using experience acquired while playing the game. The *emergence* of good play is *the* objective of Samuel's study.

What Samuel Did about It

With this in mind, let's see what Samuel did to overcome the obstacles (see Figure 4.1). I'll first describe each of his ideas briefly, then develop a more careful description, showing how each idea leads to techniques the checkersplayer uses to overcome the obstacles described.

1. The checkersplayer approaches the problem of perpetual novelty by lumping different board configurations into sets according to features they have in common.

   A typical feature of a board configuration, called "pieces ahead," is the net number of pieces the checkersplayer has in

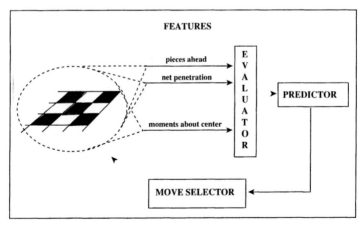

FIGURE 4.1
Overview of the checkersplayer.

excess of its opponent. A great many boards share the feature "one piece ahead," and so it is with other well-chosen features. Features, unlike the individual board configurations, appear repeatedly in game play. So a checkersplayer guided by features no longer has to deal with perpetual novelty. The essence of finding features that are helpful lies in throwing away irrelevant details, as in any modeling attempt. The features selected amount to building blocks from which to reconstruct important aspects of the boards encountered.

2. The checkersplayer learns without advice about the value of individual moves by (a) predicting the consequences of an action (move sequence), then (b) revising that action if the prediction is not verified after the action is taken.

Revision based on failed predictions seems an obvious way to learn. This approach requires no advisor whispering expert advice after each move. However, it takes considerable insight to implement the idea. Samuel's technique uses the features of end points of different short move sequences to predict the values of those end points. For example, if a series of four moves yields a board where the pieces-ahead feature jumps from 0 to 2, the move sequence is given a high predicted value. Once the predictions are made, the move leading to the best predicted outcome is executed. If the prediction comes to pass, well and good. If the prediction does *not* come to pass, then the features used to make that prediction are given less weight in future predictions. Because those features have less influence when similar situations recur, similar lines of play are less likely to be selected. The features that survive this downgrading, as experience accumulates, come to designate "good outcomes." They become subgoals sought during the play.

3. To favor a stage-setting move, Samuel again makes use of the prediction technique. In effect, he weights the features of the stage-setting move so that they predict the "good outcome" that lies farther along that path.

A subtle, stage-setting move in a game can be characterized as a move that seems to be innocuous, or even self-destructive, but is in fact the first step to some desired outcome. For example, the checkersplayer may have to "sacrifice" a piece in order to set up a later triple jump (a jump of 3 in the pieces-ahead feature). Features associated with a stage-setting move are not closely related to features that define an obviously beneficial move. By exploring lines of play from a given position ahead of time—*lookahead*—it is possible to uncover stage-setting moves. Only paths that assume good moves on the part of the opponent are of interest; otherwise the predictions involved would be precarious. (This approach, obviously, involves modeling the opponent, the technique we'll look at next.) Once a path is chosen, Samuel's program emphasizes—*weights*—the features in the starting position that predict the favorable outcome.

4. Samuel's approach to modeling the opponent is at once straightforward and subtle: assume the opponent has the same knowledge about advantageous features and moves that you do.

To implement this idea, Samuel's program assumes that the opponent has access to the same weighted features as the program itself, an unusual version of the Golden Rule. Assume opponents will do unto you what you can do unto them. If the checkersplayer has used its experience to advantage, this model of the opponent avoids overoptimistic predictions of opponent response. In the long run, this approach leads to the famous minimax strategy: minimize the maximum damage your opponent can do to you.

Self-directed modeling of the opponent is the part of Samuel's work most often overlooked when other researchers examine his ideas. Perhaps the reason for this oversight is the form of the model. It is a function formed by giving the features weights that reflect their importance. A device developed in the 1960s for pattern recognition, called a *perceptron,* was built around a similar function of weighted features. Perceptrons have quite limited capabilities (see Minsky and Papert, 1988) and are lightly regarded in AI research. Researchers may have inferred that the inadequacies of perceptrons applied as well to Samuel's modeling of the opponent, despite the success of his efforts. If so, the inference is wrong. In Samuel's work this function of weighted features plays a quite different role: it defines a *strategy*.

The ability to model the opponent's strategy is vital to the success of Samuel's checkersplayer. It serves as grist for the program's learning mill. As the checkersplayer gains experience, it improves its model of what its opponents can do. This modeling of others' strategies takes on added interest nowadays, because of our attempts to understand systems of adaptive agents—markets, ecosystems, immune systems, and the like (see Cowan et al., 1994). As in checkersplaying, these agents change their strategies on the basis of experience. The complexities that arise bear strong similarities to the problems Samuel faced, and the design of the checkersplayer suggests ways of understanding these complexities.

Our task now is to give some weight to Samuel's weights!

Valuing Moves

Just how can we specify a strategy using weighted features? Recall first that most board configurations that occur in the course of a game are unlikely to be seen again, even in a lifetime of playing checkers. (The exceptions are the boards near the beginning and end of the game.) To compensate for this perpetual novelty, Samuel devised features that characterize large numbers of boards. We've already looked at the pieces-ahead feature, where the number of pieces of the opponent is subtracted from the number of pieces of the checkersplayer. Clearly, this number can be determined for each board. All of the many boards that share the same number, say one piece ahead, are said to have the same value for this feature. All of the other features share this property: they all lump a large number of distinct boards into a common category, much as a number lumps all groups of that number of objects into a common category (see Figure 4.1). These categories, unlike the boards, recur repeatedly during plays of the game, offering a way to escape the perpetual novelty of the boards themselves.

It seems natural to look for features that are closely associated with winning plays. The pieces-ahead feature obviously has this character, and its value is easily determined for any configuration. The checkersplayer simply counts the number of checkers on its side, then subtracts the number of pieces its opponent has. At the beginning of the game the pieces-ahead feature will have value 0, because each side has an equal number of pieces. As the game progresses, the value can shift to a positive number, say +2, or a negative number, say –3. In the first case the checkersplayer has two more pieces than its opponent, in the second it is behind by three pieces.

Samuel devised a substantial array of features. They ranged from features closely associated with winning, such as "pieces ahead" or "kings ahead," to features associated with an advantage, such as "net penetration beyond the center line," to features that seemed to be "thrown in for the heck of it," such as something called "moments about the center" (the farther the pieces are from the center, the higher the moment, as when mass is moved to the rim of a flywheel). For each board, each feature has a given numerical value.

Of course, the features are not all equally important in determining the value of a board. To take this into account, Samuel *weighted* the features according to their importance. For example, the pieces-ahead feature might have a weight of 50, while the moments-about-the-center feature might have a weight of 2. Then, if a given board has a pieces-ahead value of 1 and a moments-about-the-center value of 4, the weighted values would be 50 and 8, respectively. The overall value of a board is estimated by adding together the weighted feature values for that board. If the number is large, the configuration is estimated to be valuable (on the path to a win); if the number is negative, it is a board probably best avoided. This way of assigning a single number to each board is called a *valuation function,* designated by the letter *V.*

Not all estimates are valid estimates, so it is perfectly possible that the valuation function *V* will produce misleading estimates. For the estimate to be valid, the features must be carefully selected and the weights must be appropriate. We'll see shortly how learning can improve the value estimates by manipulating the weights used in *V.*

When we discuss the checkersplayer's procedures for improving the valuation function as it plays the game, and later when we compare emergence in the checkersplayer to emergence in neural networks, it is helpful to know the exact mathematical form of this function. The features, which are the building blocks for the valuation function, are themselves functions. That is, they map the game's states *S* (board configurations) to some set of values, say the real numbers *R,*

$$v: S \rightarrow R,$$

where $v(s)$ is the value feature v assigns to board s. Samuel typically used many features, so we add an index i to v, using a standard notational convention whereby v_i designates the *i*th feature. When there are 32 features, the index runs through the numbers 1 to 32; notationally, $i = 1, 2, \ldots, 32$. If the number of features is not predetermined, we let k be the top index, and we indicate a set of k features by the notation $\{v_i: S \rightarrow R, i = 1, 2, \ldots, k\}$. Samuel's *valuation function V* is simply a weighted sum of the feature values, so that the value $V(s)$ for a board $s \in S$ is given by

$$V(s) = \sum_i w_i v_i(s),$$

where w_i is the weight applied to the *i*th feature. This value $V(s)$ is treated as a prediction of the value of the best ending that can be attained in playing from board s on.

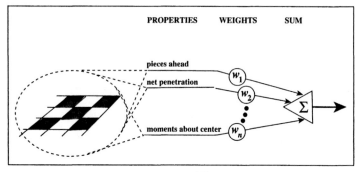

FIGURE 4.2
Checkersplayer's valuation function.

From Valuation to Strategy

Just how does this sum V determine a strategy? Recall that a strategy is any procedure that determines a unique move for each legal configuration. If we trust the value the function V assigns to each board, then at each move we pick the move (branch) that leads to the board with the highest value. That is, we move the piece that yields the highest-valued result. Under this rule, V determines a strategy, because it determines a move for each configuration.

A strategy may, of course, be good or bad. It is perfectly possible to specify a strategy that always leads to a loss. If V is a poor value predictor for certain boards, the checkersplayer may make poor moves when it selects those boards. *The whole object of Samuel's learning technique is to improve the prediction, and the associated strategy, as experiences (samples of game play) accumulate.*

A Learning Procedure

Samuel's program "learns" by using the information provided when V makes a value prediction that is later contradicted. More carefully, let $V(s)$ be the value assigned to board s; $V(s)$ is a prediction of the best ending that can be attained in playing from board *s* on. If, at some board *s'* farther along, *V(s')* is different from *V(s)*, the checkersplayer knows that something has gone astray (Figure 4.3). For example, assume that *V* assigns the value *V(s)* = +4 to board *s*, and then five moves farther along *V* assigns the value *V(s')* = –2 to board *s'* attained at that point. Clearly *V* is not being consistent in its prediction of the value that will ultimately be attained.

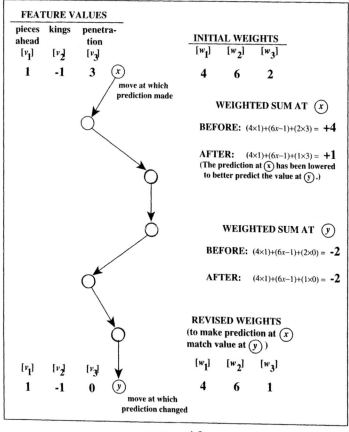

FIGURE 4.3
Prediction.

Samuel adopted the commonsense view that the outcome of the game becomes more obvious as the game gets nearer to the end. Under this view, in the example just given, the earlier estimate $V(s) = +4$ is treated as less reliable than the later value $V(s') = -2$. To compensate for this error, the checkersplayer should adjust the weights in V so that $V(s)$ yields -2, instead of $+4$, for the earlier board s. If, in the future, the checkersplayer encounters a board with the same feature values as s, it will now avoid that board (because the new weighting assigns it a negative value.)

Of course, this later estimate $V(s') = -2$ might also be in error. But eventually we come to the point where s' is the last board in the game. Then the checkersplayer gets definite information about just how close the game was. V can be adjusted to give a number that reflects the closeness of the game, with full assurance that the number is based on correct information.

If we step back from the details, we can see that the objective is to adjust V so that, after repeated plays, it makes consistent predictions throughout the game. This ideal will never be achieved in practice, but it does provide a solid guideline. It suggests that the checkersplayer continue to modify V during the play of the game, so that earlier predictions match later predictions. Under this regime V first "stabilizes" in situations s near the end of the game. Once $V(s)$ becomes an accurate predictor at s, s can actually serve the same role as s', providing a secure value for still earlier adjustments. During repeated plays of the game, earlier and earlier adjustments yield reliable predictions of the final outcome. That is, as the checkersplayer accumulates experience, the reliable predictions move to earlier and earlier parts of the move tree.

It is difficult to *prove* that this approach converges on a powerful (say minimax) strategy, and there are some further difficulties (see *Cautions* on page 67). Still, the idea is plausible and, as Samuel showed, it works in practice. The checkersplayer not only learned to beat Samuel consistently (and Samuel was a superior checkersplayer), but it also learned to play a winning game against champion players.

Making the Learning Procedure Work

We now have a general guideline that shows how the checkers-player can learn by improving the valuation function *V* as experience accumulates. But *how* is this guideline to be followed in practice?

Samuel simplifies the problem by fixing the features $\{v_i\}$ at the outset. That is, he treats the features as a fixed part of the checkersplayer's "sensory apparatus." This simplification mimics the fixed sensory apparatus of an organism, which can only sense and react to limited parts of its environment. A primate, for instance, can only see certain wavelengths of light (the visual range, but not infrared or ultraviolet), can only hear certain sounds (roughly in the range from 50 to 20,000 cycles per second), and so on. It can only react to things it can detect. Similarly, Samuel's checkersplayer can only react to situations its features $\{v_i\}$ can distinguish. This restriction puts the full burden of learning on weight changes in *V.*

If we accept this limitation (but see *Comparisons* in the next chapter), the question becomes more focused: how should the *weights* be modified? A clue to the answer surfaces when we note that, in most situations, only a handful of features will take on values greatly different from 0. A feature like pieces ahead will have value 0 in a close game, because a game usually is *not* close if one side or the other is ahead in pieces. It is the features that take on large values (positive or negative) which characterize a particular board (state).

When a prediction fails, it will be the features with large weighted values that have contributed most to the sum, so they will be most responsible for the failure. Consider a line of play for which the moments-about-the-center feature has consistently had a large value, say 3, but suddenly has a much lower value, say 0, for the current board. The moments feature contributed to an overestimation, perhaps leading the checkersplayer into a trap. The checkersplayer can compensate by reducing the weight of the moments feature. The contribution of moments to *V* at the earlier

The power of this procedure, and its dependence on the features as building blocks, is explicit in the mathematical formulation. The selection made by $V(s)$ will be little influenced by a feature $v_i(s)$ that has a value close to 0, because that feature will contribute little to the sum $\sum_i w_i v_i(s)$. Even if the weight w_i is large, the product $w_i v_i(s)$ must be close to 0 if $v_i(s)$ is close to 0 (unless w_i is extraordinarily large—not possible for the weights Samuel allowed). Because only features with larger values for s contribute to $V(s)$, it is those features that characterize s for purposes of prediction.

Samuel used these observations to determine which weights to change. Consider a board s', attained in play leading from s, for which $V(s') << V(s)$. That means the prediction made at s, by $V(s)$, is much higher than the later prediction, $V(s')$ at s', along the very path selected by V. Following Samuel's heuristic, we want to bring the earlier prediction in line with the later one by lowering $V(s)$. An obvious way to lower $V(s) = \sum_i w_i v_i(s)$ is to change the weights of the features $v_i(s)$ that have large values at s. If feature $v_i(s)$ has a large positive value, decrease its weight, thus decreasing its contribution to $V(s)$; if it has a large negative value, increase its weight, thus increasing the amount it subtracts from the sum. Each of these large-valued features will contribute less to the sum $V(s)$, so the sum is reduced. In other words, the prediction at s, $V(s)$, is devalued, bringing it more in line with the later lower value $V(s')$.

Although this approach to revising the weights is generally sound, it must be carried out with a bit of care. If the feature v_i also has a large value at s', then any change in the weight w_i will also cause a large change in $V(s') = \sum_i w_i v_i(s')$, at the later board s'. This concurrent change of $V(s)$ and $V(s')$ defeats the objective of having $V(s)$ accurately predict $V(s')$. It suggests a revision: change only the weights of features that have large values at s but small values at s'. Then weights that cause large changes in $V(s)$ will have little effect at s' because the features have small values there and contribute little to the sum (see Figure 4.3). Intuitively, it is exactly these features with large values at s and small values at s' that distinguish the situation at s from the situation at s'.

boards will be reduced, but the value of V at the current board will be little changed because the feature already has a small value there. In this way, V's prediction at the earlier boards has been brought more in line with its prediction at the current board. (A similar argument can be made for features that have large negative values early on, followed by small values at the current board.)

In general, features with large early values and small current values are *implicated* in prediction failures. The corrective action is to change the weights of the implicated features, bringing the earlier values closer to the current value. No external advice is required!

Cautions

By design, the pieces-ahead feature treats all boards at which the checkersplayer is one piece ahead as equivalent. And so it is with the other features. This procedure has the positive consequence of letting us experience the feature over and over again in a variety of situations, building up experience accordingly. It also has a negative consequence. The weight assigned to a feature can, at best, give the "average" importance of that feature over all the different situations for which the feature has a large value.

Because being one or more pieces ahead is *usually* advantageous, the weight of the pieces-ahead feature is likely to be large and positive. But there are some situations in which achieving a piece ahead is a prelude to a trap—a gambit. In those situations, other features must take large negative values to override the positive contribution of pieces ahead. In short, there must be other features that characterize the trap. Only then can the checkersplayer avoid the trap. When we discuss *default hierarchies* in Chapter 12, we'll see one version of this technique of overriding a general feature in specific situations.

There is another reason for caution, one we will encounter repeatedly in other examples of emergence. The predictions V are, at best, estimates. They are after all based on a limited sample of possibilities for the game, in terms of both play and opponents. Samuel compensates for this limitation by changing weights only enough to move the earlier prediction $V(s)$ *somewhat closer* to the later prediction $V(s')$. With this provision, no single situation can cause a large change in the weights. Instead, the small changes accumulate, producing a weight that averages over the many situations where the feature is relevant. If there is a consistent bias, the weight will accumulate enough changes to acquire a large value; otherwise it will stay close to 0.

Averaging in this way lets the weight "ignore" the occasional situation in which it makes a poor prediction. The occasional adverse experience of a trap does not change the weight of pieces ahead much from a value that is well adjusted for most other (favorable) situations.

Insufficient experience with skilled opponents can compound this difficulty. If the checkersplayer always plays poor opponents, its V will not yield a robust strategy. Human players, of course, suffer the same limitation. About the only remedy is to confront the checkersplayer with a variety of competent players. Samuel, as partial compensation, did come up with a bootstrapping technique that allowed the checkersplayer to play against itself. We'll look at that technique in the next section.

Despite these cautions and difficulties, and despite the lack of a formal proof that weight changing à la Samuel converges to a powerful strategy, the technique works well in practice. All in all, his was a remarkable achievement. It is surprising that so little effort has been devoted to finding out exactly *why* the procedure works, because weight changing is the heart of checkersplayer's emergent ability.

Emergent Consequences of Weight Changing

Even when Samuel's weight-changing technique is well understood, several consequences are not easily discerned. These can be grouped under five broad headings:

1. *Subgoals.* Though it seems that the evaluation function makes no provision for subgoals, in fact it does, providing subtle direction when there is no clear path to a win or an obvious advantage.

2. *Anticipating the opponent.* The checkersplayer must impute a strategy to the opponent if it is to anticipate the opponent's actions; the valuation function can serve as a guide to the opponent's likely responses.

3. *Toward minimax.* The valuation function only indirectly minimizes the maximum damage the opponent can inflict (minimax), yet it captures important elements of this idea.

4. *Bootstrapping.* The checkersplayer can improve its performance by playing against itself.

5. *Lookahead.* Knowing the rules of the game, the checkersplayer can look ahead several moves using its model of the other player, changing weights on the basis of anticipated outcomes.

The discussion that follows shows, for each of these topics in turn, how they emerge from the basic weight-changing technique.

Subgoals

We now know that Samuel's checkersplayer learns by using information during the play, without waiting for the payoff at the end and without requiring expert advice along the way. At first sight it seems to work without subgoals. How can this be, if subgoals are a critical intermediary when there is no clear path to the long-term goal?

The answer, as one might suspect, is that the checkersplayer makes extensive use of subgoals; they are just not labeled as such. A closer look reveals some interesting subtleties of the checkersplayer, pointing up the learned nature of its subgoals.

Let's start by looking again at the pieces-ahead feature. That feature gains a high *weight* under the learning procedure, because it is highly correlated with winning. On the other hand, pieces ahead has a *value* near 0 in a close game, because a game is not close if one side or the other is several pieces ahead. Thus, pieces ahead will have little influence on the choice at board s, if all moves from s lead to close positions. What then?

It is then that the features not so closely correlated with winning come into play. An oddball feature, like moments about the center, will likely have a nonzero value. In a close game this feature, despite its small weight, will influence the sum V more than the obvious features, which have values close to 0. When the game is close, the strategy imposed by V accordingly selects boards rated by oddball features having nonzero values. That is, those features determine (characterize) the selection when the obvious features give no information. In this sense, the oddball features designate the subgoals, determining play when the way to the goal is not obvious.

The checkersplayer must learn which oddball features serve it well when the more obvious features supply no direction. The same averaging and compensatory procedures that were at work for the more obvious features are also at work here. The weights assigned to oddball features tend to be smaller, because those features are less directly correlated with successful outcomes. Some will prove completely unreliable, ending with weights close to 0 and having little influence on the play. Others will encapsulate some important piece of information, characterizing good directions in close situations. It is the latter features that acquire weights different from 0, and it is these features that V seeks out when the more obvious features do not discriminate among options.

Anticipating the Opponent

Now we come to the problem of modeling the checkersplayer's opponents. Because the checkersplayer faces different opponents, the same plays on its part do not always lead to the same outcome. However, by concentrating on the bad outcomes—the minimax approach—the checkersplayer can partially alleviate this difficulty.

An undesirable outcome can only occur because (a) the checkersplayer has made a bad play, or (b) the opponent has made a good play. In either case, the checkersplayer is well advised to assign a low (possibly negative) value to configurations that lie in that direction. On the other hand, a desirable outcome presents an ambiguous situation. The desirable outcome may occur because of good play on the part of the checkersplayer, but it can also occur because of poor play on the part of the opponent. If the outcome is the result of an opponent's poor play, then it is unwise to make any adjustments. That line of play is unlikely to recur, either because the opponent has learned or because a different opponent plays a better game.

This consideration suggests that the checkersplayer should concentrate on avoiding bad outcomes, rather than directly seek good outcomes. Bad outcomes should be avoided whatever the cause, while good outcomes may not hold up against other opponents. This approach can be implemented by only adjusting the weights in V when its prediction is too high relative to later predictions. As AI chess expert Berliner said in 1978, the secret of good play is avoiding big mistakes.

This revision completes the rules for revising the weights. We apply the earlier heuristic for revising the weights in $V(s)$ only when some later board s', along the path selected by V, exhibits a much *lower* value. That is, the weights in V are only revised when $V(s)$ is much greater than $V(s')$; the opposite case, $V(s)$ much smaller than $V(s')$, is ignored for the reasons just mentioned.

Toward Minimax

Just how does this regime tend toward a minimax approach to checkers? It would be satisfying to be able to prove that this approach converges to a minimax strategy, given enough time and an appropriate set of features. Alas, as already mentioned, no such proof exists. The best we can do is give a plausible argument that local actions are "minimax-like."

The main argument has already been made. Predictions can be "moved back," stage by stage, from secure predictions at the end of the game to ever-earlier stages. In this process, V is continually adjusted to assign low values to boards preceding situations that have unexpectedly low values. It follows that the strategy defined by V avoids predicted damage. Moreover, features assigned large negative weights, because of a very bad outcome, are likely to be avoided in other situations in which that weighted feature has a negative value. This is consonant with Berliner's advice about avoiding big mistakes.

As experience accumulates, the choices in more and more situations will be determined by such avoidance. That is, the highest-valued alternative will be the one that has (so far) not led to a bad outcome. The "surviving" alternatives will be the ones selected because the strategy defined by V selects, on each move, the legal move leading to the board of highest V value. In this way the checkersplayer comes to the minimax maxim, adapted to *estimated* outcomes: minimize the maximum *estimated* damage your opponent can do to you.

Bootstrapping

Samuel's procedure for modifying the weights does its minimaxing job well enough that the checkersplayer can bootstrap itself into good play. To do this, a version with fixed weights is used as a sparring partner for the learning version just discussed. After a while, the learning version consistently beats its fixed-weight sparring partner. Then the learning version's weights are transferred to the sparring partner, and the process is repeated.

Under this regime, the learning version confronts an increasingly sophisticated opponent. New lines of play are explored, and the learning version continues to increase its competence. Because the object is to find and avoid the damage that can be inflicted by a capable opponent, this bootstrapping procedure yields a strategy that provides insurance against gambits, traps, and other lines of play that start out looking promising and end up causing trouble.

Lookahead

We have thus far discussed only one way to use the evaluation function V: at each decision point during the play of the game, the choice of the next move is made on the basis of V's prediction. As play proceeds, the weights in V are changed each time there is a substantial change (for the worse) in V's prediction. Another way to use V involves a technique called *lookahead*.

With lookahead, the checkersplayer uses the rules of the game to generate a part of the tree of moves starting from the current position (see Figure 4.4). It first generates the boards that can be attained from the current position, then it generates the boards that can be attained from those boards, and so on for several layers. That is, the current position is the root of the lookahead tree, and the program generates as much of the tree as it feasibly can in the time available to it. We know that this tree bushes out very rapidly, so the checkersplayer could only investigate *all* possible lines of play to a depth of five or six moves at the time Samuel was testing his program.

After the checkersplayer generates the lookahead tree, it applies the function V to the terminations of the tree generated—the leaves of the lookahead tree. In this way, the checkersplayer evaluates a range of future possibilities without actually executing a line of play. In particular, the program evaluates the result of *all* the possible responses of its opponent over the limited lookahead span. With this information, the checkersplayer can again use the heuristic that later evaluations should be trusted more than earlier ones.

Starting from the leaves of the lookahead tree, the checkersplayer works backward layer by layer to the current position, the root of the lookahead tree. In those layers where the opponent has the move, the checkersplayer assumes, as before, that the opponent makes the move that is most disadvantageous to the checkersplayer. To make this decision, the checkersplayer assumes that the opponent has exactly the same knowledge that it has itself—it assumes that the opponent knows V, with its features and weights.

Accordingly, the checkersplayer assumes that the opponent always selects the alternative with the *lowest* value of V because in checkers what is bad for the checkersplayer is good for the opponent, and vice versa. In short, Samuel uses V to *model* the oppo-

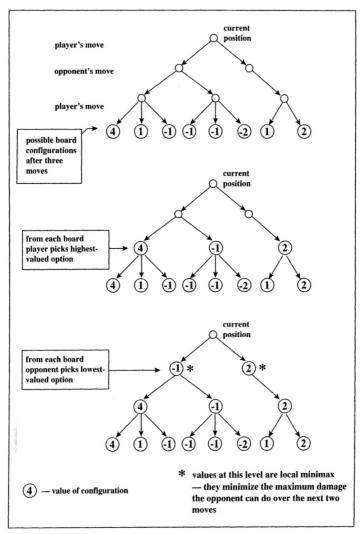

FIGURE 4.4
Lookahead and minimax.

nent's preferences, as well as to determine the checkersplayer's strategy. The result of this procedure is to select a line of play that is *locally minimax* according to the evaluation function V (see Figure 4.4): the checkersplayer minimizes the maximum damage predicted by V within the lookahead tree.

With this provision, lookahead allows the checkersplayer to compare the value assigned directly to the lookahead tree's *root r* to the value of the *leaf l* belonging to the line of play chosen by the minimax procedure (see Figure 4.4). It changes the weights in V, using the technique just discussed for actual play, so that $V(r)$ more closely predicts $V(l)$. In this way the checkersplayer can change its weights *before* it executes a line of play.

Summing Up

It was a stroke of genius on the part of Samuel to use his sum of weighted features, V, as an internal model of the opponent. With this one stroke he (a) greatly amplifies V's power as a predictor, (b) makes possible rapid, sophisticated learning in the absence of reinforcement, (c) uses model-based anticipation (lookahead) as a basis for learning, and (d) nicely implements Berliner's exhortation to avoid big mistakes. In addition, there is reason to believe that the resulting strategy is locally minimax.

We are nowhere near exploiting these lessons that Samuel put before us almost half a century ago. His checkersplayer, despite the parochial setting, clearly relates learning and anticipation-based actions to emergent capabilities. It is also a paradigmatic example of self-improvement through the use of an internal model for "introspection." Most important, for our purposes, the checkersplayer teaches lessons broadly applicable to the whole study of emergence. These lessons can be grouped under four headings:

1. *Generated complexity.* The checkersplayer clearly illustrates the organized complexity that can result from simple rules and procedures.

2. *Learning without immediate reward.* Prediction makes improvement possible, even when there is no referee to distinguish "right" from "wrong."
3. *Credit to stage-setting actions.* Learning in complex environments requires a defined procedure for discerning early actions that set up later, obviously good responses.
4. *Modeling other agents.* In multiagent environments, emergent phenomena often arise from anticipation of the actions of other agents.

By examining these lessons with emergence in mind, we can preview some of the broader issues to come.

Generated Complexity

First of all, we learn that even a game as simply defined as checkers provides an almost inexhaustible variety of settings (board configurations). This ability to generate extreme complexity from simple specifications is daunting because it assures that complexity will be pervasive in the world around us. At the same time, it gives hope that we can find simple rule-governed models of that complexity. Generated complexity is essential to emergence, and we will have to look at it closely if we are to understand emergence.

The perpetual novelty generated by simple specifications supplies a strong rationale for constructing and examining models. In a milieu such as checkers, we can only "see the same thing twice" by ignoring many of the details of each setting. By eliminating details, we gain the repetition that enables us to build a repertoire of learned actions of the form "IF the situation shows feature(s) *X*, THEN take action(s) *Y.*" Deciding what is salient and what is detail is, of course, a matter of taste and experience. Therein lies the art of model building.

Sensitivity to salience and detail touches on an ability that seems effortless in humans and other mammals: when confronted with a complex scene, in the countryside or in the city, we can quickly parse it into familiar elements. "There are three trees, a house, a cow, . . . " or "There are five skyscrapers in a row, two streetlights, a fire hydrant, . . . " It is a process that takes almost no conscious effort. Nonetheless, it is difficult to capture this ability in computer-based procedures—difficult to the point that we still have no programs that achieve the task flexibly.

In fact, this "effortless" parsing is a remarkably subtle process. The scene is parsed into *reusable* building blocks, not just arbitrary components. We see trees over and over again, though we never see even the same tree in the same way. Differing light and differing angle provide a new impression on the eye's retina each time the tree is seen. Still, by dropping details, we see trees in all sorts of contexts, and in wonderful variety. More remarkably, we readily parse the tree further into "roots," "trunk," "branches," and "leaves." These building blocks, put together in different ways, let us model and recognize different kinds of trees. They act much like a child's building blocks, because they can be combined in many ways. As in games, only certain combinations are legal: unless we are trying to construct a "baobab," roots are at the bottom and leaves are at the top. Sometimes, for comic or startling effects, we try "illegal" arrangements, but those are the exception, not the rule.

We can go on to use the tree model in a metaphorical sense, transferring some of our understanding of trees to more remote contexts. We've talked about game trees, even using the notions of root and leaf to extend our comprehension. In Chapter 11 we'll look more closely at metaphorical usage and its relation to models.

Learning without Immediate Reward

Samuel's insight about getting along without a referee comes in two parts. The first is the realization that failed predictions can serve as well as overt reward as a basis for improvement. In games, and in most realistic situations, reward occurs only after an extended sequence of actions, and it provides only a small amount of information. The reward does *not* pinpoint the critical acts along the way. Adjustments made only on the basis of reward lose information acquired en route.

Predictions made along the way, on the other hand, make it possible to use interim information to improve performance. Samuel implemented this insight by treating the valuation function $V(s)$ as a prediction of the value that could be attained by using the strategy defined by V from point s on. He treated each board encountered as a root of the subtree, with V serving to estimate the value that could be attained by an appropriate sequence of moves within that subtree. Whenever subsequent events (or lookahead) show the prediction at board s to be overly optimistic, the checkersplayer revises the weights of features that enter most strongly into giving $V(s)$ too large a value.

The second part of Samuel's insight follows from an observation about minimax. If both the checkersplayer and the opponent consistently follow the strategy of minimizing maximum damage, the same value is assigned to each board along that path. If V is treated as an approximation to minimax, it should satisfy a similar criterion: its predictions should be the same at every stage of the play. Accordingly, V should be revised so that its predictions are consistently the same, always giving a later value precedence in the revision procedure. This leads to a local, V-based minimax criterion.

Credit to Stage-Setting Actions

Good play in a board game comes from subtle, stage-setting moves (perhaps sacrifice of a piece) that make possible later, obviously good moves (such as a triple jump in checkers). Even a weakly adapted V favors a triple jump when it is offered because of the easily learned, high weight on the pieces-ahead feature. The trick is to provide appropriate weights to features that influence the choices that set the stage.

Stage-setting has a crucial role when the game is close. A game is close when the features strongly associated with winning (pieces ahead and kings ahead in the case of checkers) have values close to 0. That is, the obvious features exert no influence on V in close situations—they have the value 0 for all local options. The decision then falls to the more arcane features, which differ in value over the local options. In effect, these features define subgoals to be sought in the absence of an obvious direction. Their weights must come to favor choices leading to later, obvious advantage.

Samuel's learning procedure automatically provides credit to prescient stage-setting choices. When a choice based on V's prediction leads to a disadvantageous situation—that is, if the prediction is not fulfilled—the weights are changed so that choice is avoided in the future. Under repetition, the only "survivors" are weights that avoid choices leading to unfavorable consequences. The corresponding weighted features, if they have positive values, are "selected" as subgoals to be sought. Contrariwise, weighted features with negative values become signs of disaster to be avoided. In both cases, V gains the ability to favor useful stage-setting choices.

Modeling Other Agents

To model the opponent, Samuel boldly assumes that the opponent knows everything the checkersplayer knows. The checkersplayer acts on the assumption that the opponent will exploit the strategy defined by V wherever possible. Because checkers is a zero-sum game (whatever the checkersplayer gains, the opponent loses, and vice versa), predicted losses under V are gains for the opponent, and vice versa. V "turned on its head" thereby models the opponent's strategy. This leads back to the use of the minimax concept in lookahead. In "backing up" from the leaves at the bottom of the lookahead tree, Samuel assumes that the opponent selects moves that *minimize* V, over the options available. V, combined with V "on its head," gives an estimate of the likely line of play.

Of course, the opponent may not really have a strategy that mirrors V. If the opponent's strategy is less effective than V, this pessimistic minimax outlook will simply lead to a better outcome than predicted. When this happens, the weights are not altered, because another opponent will be unlikely to make the same mistakes. On the other hand, the opponent's strategy may be superior to V in some ways, yielding outcomes unanticipated by V. The result is a learning opportunity. The weights in V are altered to avoid these unanticipated disasters. V then attributes more skill to its opponents, and its strategy improves in terms of minimizing the damage the opponent can inflict.

Although it is easy to devise flaws and counterexamples when modeling the opponent in this way, the pragmatic point is that the system works in practice. In complicated situations involving many agents that interact and learn or adapt, the efficacy of this simple modeling technique provides a useful starting point.

CHAPTER 5

Neural Nets

On FIRST THOUGHT, modeling networks of neurons would seem to be an enterprise having little in common with modeling a checkersplayer. My own first reaction to Art Samuel's checkersplayer, as I mentioned earlier, was to think the ideas fascinating but far removed from the study of neural networks. The previous chapter established Samuel's insights as much deeper, and more far-ranging, than conveyed by the word "fascinating." Now I must rescind the "far removed" as well. At the right level of comparison, the models of neural networks have much in common with Samuel's checkersplayer, as we'll soon see. We'll also see that a close look at neural nets takes us further in our understanding of the phenomenon of emergence. We can preview the new perspective by retelling Hofstadter's suggestive (1979) metaphor of the ant colony.

Individual ants are remarkably automatic (reflex driven). Most of their behavior can be described in terms of the invocation of one or more of about a dozen rules of the form "grasp object with mandibles," "follow a pheromone trail (scents that encode 'this way to food,' 'this way to combat,' and so on) in the direction of an increasing (decreasing) gradient," "test any moving object for 'colony member' scent," and so on. (To actually perform a computer simulation of an ant following these rules, the description of the rules would have to be somewhat more detailed, but these phrases give the gist.) This repertoire, though small, is continually invoked as the ant moves through its changing environment. The individual ant is at high risk whenever it encounters situations not covered by the rules. Most ants, worker ants in particular, survive at most a few weeks before succumbing to some situation not covered by the rules.

The activity of an ant colony is totally defined by the activities and interactions of its constituent ants. Yet the colony exhibits a flexibility that goes far beyond the capabilities of its individual constituents. It is aware of and reacts to food, enemies, floods, and many other phenomena, over a large area; it reaches out over long distances to modify its surroundings in ways that benefit the colony; and it has a life-span orders of magnitude longer than that of its constituents (though for some species the life-span of the queen may approximate the life-span of the colony). To understand the ant, we must understand how this persistent, adaptive organization emerges from the interactions of its numerous constituents.

Like the ant colony, the central nervous system (CNS) is composed of numerous interacting individuals, called *neurons*. Individual neurons, like individual ants, have a behavioral repertoire that can be reasonably approximated with the help of a small number of rules. And, like the ant colony, the behaviors mediated by the CNS are much more complex, in both time and space, than the behaviors of the constituent neurons. There are, of course, important differences between an ant colony and the CNS. For example, the interconnections and interactions of the neurons are largely "wired" in place, whereas the ants have a fluid, shifting network of interactions. Still, the central mystery is much the same in both cases: how does a persistent, flexible organization emerge from relatively inflexible components?

Because the details of an ant colony are more observable than the details of the CNS—we can see the individual ants and observe their interactions with the naked eye—emergence seems somehow less mysterious. In this respect the simile is helpful, because it makes it plausible that a neural network can acquire a repertoire of behaviors that far exceeds the repertoire of its constituent neurons. In building models of such networks, we hope to examine this phenomenon in much the same way as we might observe an ant colony. We can perturb the simulated network in different ways, sorting out the properties of the individual neurons that have a key role in the emergence of organization. Further, because the network provides a different structure than Samuel's valuation function, we can examine facets of emergence not readily observable in Samuel's checkersplayer, while examining his broad insights in a different context.

Some Facts about Neurons

The most striking fact about the human CNS is its sheer magnitude. Even the most complex human artifacts, such as digital computers, are orders of magnitude simpler. The possibilities for interaction in the human CNS, which has about 50 billion neurons, are roughly measured by *fanout*, the number of direct contacts a neuron makes with other neurons in the system. Typical CNS neurons have a fanout ranging from 1,000 to 10,000, whereas typical active elements in a digital computer have a fanout of less than 10! This is a difference of about three orders of magnitude. In the sciences, three orders of magnitude (the jump between atomic nuclei and the orbits of electrons, for instance) is enough to call for a new science (from atomic physics to chemistry). Our experience of artifacts with fanout less than 10 does little to prepare us for the complexities of systems with fanout in excess of 1,000.

Descriptions of neurons outside the technical literature rarely give much attention to the intricacies of these cells. It will not repay us here to go deeply into the chemistry and physiology of neurons, but it *is* worthwhile to sketch some of the major features.

First of all, a neuron is a cell with some intricate extensions of its surface (see Figure 5.1). One type of extension, called the *axon*, is much like a tree, starting out from the surface in a single trunk, then branching progressively as it gets farther from the cell body. Axons may sometimes have great length, providing long-distance connections to other neurons in the brain. It is the axon's branches that provide the cell's fanout. The other extensions,

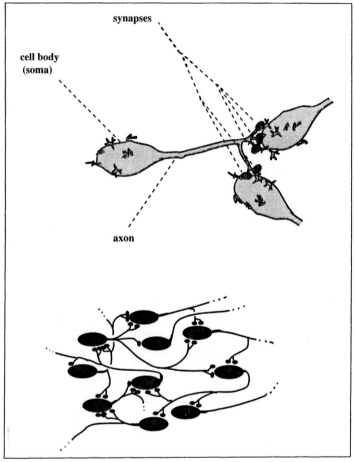

FIGURE 5.1
Neuron and neural net (highly stylized).

called *dendrites,* are usually bushier and typically do not extend far beyond the cell's body. In this discussion I concentrate on the axons, as the main determiners of interaction, though dendrites certainly have an important role. Even with this simplification, the picture of neural activity developed here goes considerably beyond that in a typical discussion of artificial neurons.

The axon's branches terminate at the surface of other neurons. These points of contact are called *synapses,* and the end points of the branching axon are actually separated from the body of the neuron contacted by a very thin space called the *synaptic gap.* This gap is so thin (about 100 angstroms) that chemicals diffuse across it in a matter of microseconds.

Neurons interact via pulses of energy that propagate along their axons. Because energy is supplied locally along the axon, the propagation of a pulse is much like the burning of a fuse—when the pulse comes to branches in the axon, it splits without attenuation. As a result, the pulses arriving at the axon's synapses are the same size as the pulses that started out at the cell body, despite the profuse branching of the axon (see Figure 5.1). Moreover, pulses throughout the CNS are pretty much the same size, so that the amplitude carries little information beyond the presence or absence of the pulse.

When the pulse arrives at a synapse, it causes the release of biochemicals called *transmitters,* which diffuse across the synaptic gap. If enough pulses arrive at the surface of a neuron over a short interval of time, the neuron *fires,* propagating a new pulse down its own axon. In effect, the neuron adds up the incoming pulses and if there are enough of them, it sends out a pulse signaling this fact. The time it takes the body of the neuron to integrate this incoming information is long with respect to the time it takes a pulse to propagate between neurons. So this processing time sets the rate at which the clock "ticks," and we can, to a good approximation, model the CNS with time-steps mimicking this rate of ticking.

The release of chemicals at a synapse can be more or less efficient, depending on the past history of pulses passing across the gap—much as exercise can improve muscle efficiency, while disuse can decrease it. This effect of use and disuse was early hypothesized as a basis for learning (see Hebb's 1949 work), and the changes theorized have been progressively confirmed with the increase in our sophistication in tracing biochemical activity. If we take this effect into account, we can think of the synapses as weighted according to past experience, much like the weighted features in Samuel's checkersplayer.

Modeling Neurons

To construct a useful model of the CNS using this information, we begin with the idea that the activity of the CNS is described by which neurons are firing during a given time-step. Because of the uniformity of the pulses emitted, we can simply say that each neuron is either "on" (emitting a pulse) or "off" (quiescent) during a time-step. We can further simplify the model by assuming that all pulses arriving *within* a time-step have equal effect in determining whether or not the neuron fires (in technical jargon, we ignore the *phase* of the pulses).

We can also see a connection between these simplified neurons and Samuel's work (see Figure 5.2). This connection shows up most clearly if we return to the functional notation used to describe Samuel's valuation function,

$$V(s) = \sum_i w_i \, v_i(s).$$

First, consider the $v_i(s)$. For Samuel, they are the inputs to the valuation function supplied by the feature recognition procedures. For the neuron, we can consider the $v_i(s)$ to be the signals present at the synapses on the neuron's surface. If the time-step is based on the neuron's processing time, then $v_i(s)$ simply indicates the presence or absence of a pulse at synapse i. That is, the $v_i(s)$ can be thought of as taking just the values 1 (pulse) and 0 (no pulse). If we use a longer time-step, the $v_i(s)$ can be taken as numbers giving the *rate* at which pulses are currently arriving at synapse i. That is, under the longer time-step, $v_i(s)$ gives the *firing rate* of the neuron sending pulses to synapse i.

Under either interpretation, the weight w_i represents the efficiency of synapse i in sending transmitter substance across the synaptic gap. If the sum of the pulses arriving over the synapses exceeds the fixed threshold of the receiving neuron, that neuron fires in turn. (Under the firing-rate interpretation, the firing rate of the receiving neuron is determined as a simple function of the difference $\sum_i w_i \, v_i(s) - T$.)

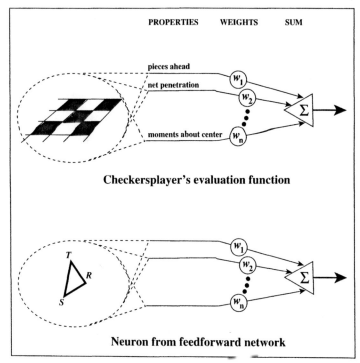

| PROPERTIES | WEIGHTS | SUM |

Checkersplayer's evaluation function

Neuron from feedforward network

FIGURE 5.2
Checkersplayer compared to artificial neuron.

We can go one more step by assuming that each neuron will fire at the end of a time-step only if the number of pulses arriving at the neuron's synapses exceeds a fixed threshold. This emphatically is *not* the case for real neurons, as we'll see in the next section. Nevertheless, this simplification is the foundation of an early, seminal paper by McCulloch and Pitts (see Kleene, 1951)—a paper that comes close to Turing's 1937 paper in the breadth of its influence on the development of computer science. The paper by McCulloch and Pitts not only laid the foundation for later formal work on neural nets, it also provided a key for later studies of computer languages (through Kleene's 1951 rewrite) and even natural languages (through Chomsky's 1957 adaptations). This same fixed-threshold simplification plays a central role in the recent resurgence of interest in artificial neural nets.

In short, both Samuel's valuation function and these artificial neurons use a weighted sum to make "decisions." And, in both cases, learning proceeds by changing the weights, though the algorithm for so doing is different in the two instances.

Networks of Fixed Threshold Neurons

Networks of fixed threshold neurons have learned to distinguish a variety of complex patterns; faces, handwriting, spoken words, sonar signals, and stock market fluctuations have all served as grist for such networks. The presenter *does* have to position the patterns carefully (centering them and putting them in a standard orientation), but the performance is impressive nonetheless. Our object now is to see how this is done and what the limitations are.

To accomplish pattern recognition, the network is set up in layers: an input layer, some interior layers, and an output layer. In this simple organization, a *feedforward* network, neurons in one layer cause neurons in the next layer to fire. The object is to have specific neurons in the output layer fire whenever the pattern to be recognized is presented to the input layer.

In the *input layer* each neuron responds to some small element in the *environment* (the scene or waveform being presented for recognition, say a triangle). For example, a picture can be divided into a large number of small squares, *pixels* (from "picture elements"), each of which is either white or black (see Figure 5.3). One input neuron is assigned to each pixel, and it fires when the pixel is black. Neurons that fire in the input layer, in response to black pixels, then cause neurons in the next layer to fire. And so it goes until the pulses reach the output layer.

Neurons in the input layer contact neurons in the next layer via their axons, synapsing with them. Neurons in that next layer fire if they are contacted by enough neurons that have fired in the first

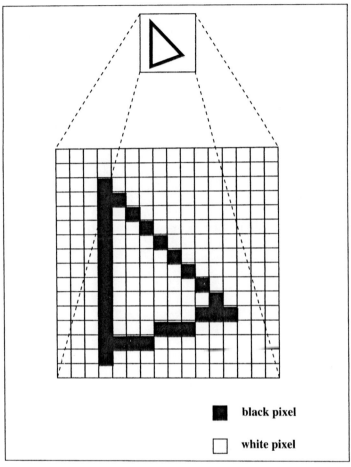

black pixel

white pixel

FIGURE 5.3
Pixels representing a triangle.

layer. The firing neurons in this layer cause neurons in the next layer to fire, and so on until the pulses reach the output layer. In the simplest cases, one specified neuron in the output layer fires if the pattern to be recognized (a triangle) has been presented. (More complicated "coded" output firings can be set up if the network is to distinguish many patterns.) When the specified neuron fires, we say the network has recognized the pattern—it has "seen" the pattern in its environment.

Of course, the output neuron may fire in error when the pattern is not present, or it may fail to fire when the pattern is present. Learning enters at this point. Feedforward networks learn with the help of a referee who "points" to the output neuron (or coding) that should have fired. If we adopt Samuel's outlook, treating the firing of an output as a prediction, then the referee designates the prediction as false when the wrong output neuron fires. For false predictions, weights throughout the network are adjusted so that the correct response will be given for future presentations of the same input. The changes are accomplished with a procedure reminiscent of Samuel's procedure for revising weights, though it differs in detail.

The rationale for weight changing in neural nets is also similar to Samuel's rationale. Weight changing in Samuel's case improves the checkersplayer's ability to recognize features that characterize appropriate moves. Weight changing for feedforward neural nets adjusts neuron firing to improve the response of output neurons to designated input patterns.

Differences and Limitations

The main difference between Samuel's checkersplayer and feedforward nets is the complexity of the decision procedure. Samuel posits a single executive, the valuation function V, whereas in feedforward nets the decision procedure is distributed among many interacting agents, the neurons. The artificial neurons with their weighted synapses are much simpler than the valuation function with its weighted feature detectors. The synapse simply records the presence or absence of a pulse (or a firing rate), while the feature detectors record the outputs of complex algorithms. Because of its distributed nature and the simplicity of its elements, a neural net is a satisfactory vehicle for studying emergence: sophisticated behaviors emerge from the aggregation of simple actions.

It is useful to distinguish the emergence observed in the checkersplayer from emergence in feedforward neural networks. The checkersplayer is making long sequences of decisions, with information about performance coming only at the end of each game. The strategy that emerges (set by the valuation function V) is a response to the complex effects of local decisions (moves). No referee is acting during the play. On the other hand, feedforward nets respond directly, instant by instant, to the patterns presented, and there is a referee that indicates whether or not the correct response has been given.

Feedforward (layered) networks have a major limitation: pulses move right through the net. If the network is n layers deep, the input pulses propagate through it in n time-steps. Once they have propagated through, they are no longer available for further processing; that particular configuration is lost to the net's memory. This lack of long-term memory for past occurrences contrasts with the memory capabilities of real neural networks. We'll see that networks with loops in their interconnections (see Figure 5.1) do better at providing long-term memory. Such networks can remember past stimulus configurations for an indefinite period, so-called *indefinite memory*. We'll come back to indefinite memory after we take a closer look at real neurons.

More Facts about Neurons

To this point, we've used a model of the neuron that is barely a caricature of the real neuron, and we've restricted ourselves to feedforward networks, which are not at all like real neural networks. Though that caricature has been productive, both in providing part of computer science's foundations and in Machine Learning studies, it is quite limiting in other respects. To see this, we require a better understanding of real neurons. In this section, I'll outline some of the salient, well-established characteristics of real neurons. In the next section, I'll show that these added characteristics bear directly on the behavior of networks with loops in their interconnections.

Time-Varying Thresholds

The first and most important added characteristic is that neurons do *not* operate with the fixed threshold of the model neurons we've so far used. A real neuron, once fired, has a period of time

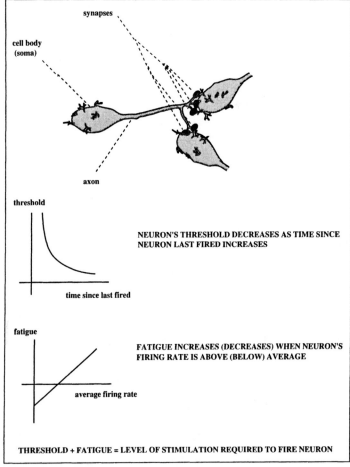

FIGURE 5.4
Additional properties of neurons.

when it cannot be fired at all—the *absolute refractory period* (see Figure 5.4), which lasts for several milliseconds. At the end of this time, the neuron becomes progressively more sensitive to incoming pulses over a period of several tens of milliseconds.

If we use the same clock rate discussed earlier, approximately one millisecond for the processing of a pulse, then a neuron cannot be fired at all for several time-steps after it emits a pulse. The rate at which the neuron can fire is thereby limited. Under very strong stimulation at its strong positive synapses (many pulses arriving within a single time-step), the threshold may be exceeded every five time-steps or so. This yields a firing rate of 200 pulses per second, about the highest observed in normal physiological conditions. If the stimulation is not strong enough to fire the neuron at that rate, the threshold decreases to the point that the neuron responds to fewer incoming pulses, yielding more typical physiological rates of 20 to 100 pulses per second.

When the connections in a network form loops, pulse rates in the network can build up to very high levels. As pulses circulating in the loops add to pulses stimulated by the environment, more and more neurons become entrained. If this continues, the network goes into an "epileptic fit," where all neurons are firing at a very high rate. A network in this state can no longer respond to its environment. Feedforward networks avoid this positive feedback by not allowing pulses to recirculate, at the cost of eschewing long-term memory of past events. For networks with loops, with their increased capacities, the time-varying threshold helps contain the buildup of pulses. By periodically putting the neurons in a state in which they cannot fire, the buildup is damped down. This, in conjunction with fatigue, gives nets with loops an insurance policy against "epilepsy."

Fatigue

Neural cells, like muscle cells, exhibit fatigue. If a neuron continues to fire at a high rate for several seconds, the waste products of its metabolism begin to accumulate. This process effectively raises the neuron's threshold, making it more difficult to sustain the high firing rate. The longer the high firing rate persists, the more the threshold increases. A negative feedback results, which eventually forces the firing to decrease. The effect also works in reverse when a neuron fires at an abnormally low rate. Then the scavengers for metabolic waste products effectively clear waste products to below-average levels. The neuron becomes hyperrecovered, more ready than the average neuron to fire.

It might seem that fatigue is incidental to the main activity, but that is not the case. Fatigue acts in conjunction with time-varying thresholds to prevent extended reverberatory firing patterns that interfere with the network's information processing. Consider a few hundred neurons connected by interlocking loops of axons and synapses. It is possible for the firing of these neurons to get into a kind of lockstep pattern, so that some subset of neurons is always ready to fire while the others are recovering. Once this synchronization occurs, the time-varying threshold is simply coerced into maintaining the pattern. The resulting reverberation could go on indefinitely, effectively removing that set of neurons from further information processing. Fatigue prevents this by gradually raising the thresholds to the point where the neurons can no longer fire at a rate that preserves the synchrony.

Changes in Synapse Weights

In feedforward net models, synapse weights are changed in response to a referee's evaluation; real synapses change in response to local events. The simplest version of such a precept is *Hebb's rule* (1949). It applies to a pair of neurons S and R, where the axon of neuron S forms a synapse at the surface of neuron R. Under Hebb's rule, if S sends a pulse at time t and R fires at time $t + 1$, then that synapse becomes more effective in firing R in the future. If we think of the synapse's effectiveness as being specified by a weight, then that weight increases.

At the time Hebb proposed his rule, it was an interesting conjecture with very little experimental evidence. But the intervening fifty years have supplied a steady accumulation of evidence that the rule is a useful approximation to some very complex bio-molecular events. This evidence centers on the production of *transmitter molecules* at synapses. Transmitter molecules diffuse across the synaptic gap to the surface of the receiving neuron; if enough transmitter molecules from enough synapses accumulate at the surface of a neuron, that neuron fires. In so doing, it effectively removes the molecules from the synaptic gap. If this happens repeatedly, the synapse increases its ability to produce the trans-mitter molecules, much as a muscle strengthens through exercise. The synapse has increased its effectiveness (weight) in a way that satisfies Hebb's law.

Not long after Hebb posed his rule, Peter Milner modified it to provide for *inhibition* (see Rochester et al., 1956). His modification allows for the fact that transmitter molecules exist that actually de-crease the probability that the receiving neuron will fire. We can think of that synapse as carrying a negative weight. It decreases the effect of the other pulses that facilitate transmission, in much the same way that a weighted feature with a negative value decreases the sum in Samuel's valuation function. Hebb's rule, so amended, has the following additional clause: if S sends a pulse at time t and R does *not* fire at time $t + 1$, then that synapse becomes *less* effective in firing R in the future. The synaptic weight may go from positive right through 0 to a negative value, if the failures are frequent.

Networks with Cycles

We've known since the time of Ramón y Cajal, the great neuroanatomist who flourished early in this century, that neurons in the CNS form a highly intertwined network with great numbers of internal *cycles* (loops, reentrant connections—see Sholl, 1956). It is unlikely that we will understand the workings of the CNS if we do not understand the effects of cycles on network behavior.

Cycles, as we will see, have a profound effect, making it possible for model networks to retain memories of the indefinite past.

Small groups of neurons can sustain circulating pulses, the reverberations discussed in the previous section. Hebb pointed out in 1949 that synapse changes allow these reverberations to change in response to experience. The resulting groupings Hebb calls *cell assemblies*. They amount to memories, shorn of detail, of stimuli that have proved salient in the networks' experience. These cell assemblies serve as building blocks from which to recreate more complex memories. They also make it possible to anticipate the future, much like the lookahead in Samuel's checkersplayer. In short, networks with cycles can generate behaviors that far surpass the limited pattern recognition capabilities of feedforward networks.

Indefinite Memory

Networks with cycles exhibit a phenomenon, *indefinite memory*, that completely distinguishes them from feedforward networks. Indefinite memory is basically the ability to remember events that have occurred in the indefinite past. Roughly, it is the ability to say that one has seen three comets in the past, no matter how much time has elapsed. Indefinite memory *cannot* be achieved by feedforward networks. It is a technical topic, but the effort to understand it is well repaid.

To define indefinite memory carefully, we have to delve into models that McCulloch and Pitts introduced, the world of "logical" neurons. First, we set up a feedforward network of logical neurons that captures the notion of the *"exclusive or"* (sometimes called *inequivalence*). This network, with two inputs and one output, produces an output pulse only if there is a pulse on exactly one of its inputs. (Whence the name—the output occurs only when the two inputs are not equivalent.) Next, we connect the output of the *exclusive or* to one of its inputs, obtaining a network with one cycle (loop) that feeds output pulses back to the input (see heavy solid line in Figure 5.6, page 100). The behavior of this one-cycle network gives a simple example of indefinite memory.

The effect of the cycle is striking. The network, which now has only one free input, can remember whether or not the number of pulses on that input wire, *from the time the network was started up,* has been even or odd. The network memory extends into the indefinite past. Stated another way, this network counts pulses, retaining the parity of the count (even or odd). Transistorized versions of this network provide both the basic bit-counting circuits (arithme-

---

The formal definition of the *exclusive or* network starts with an artificial neuron having just two inputs and a fixed threshold of +1 (see Figure 5.5 a, page 99). For this neuron, the first input I_1 carries weight +1 and the second input I_2 carries weight −1. The behavior of this neuron is best specified by a small table (see Figure 5.5 b). We can see from that table that when input I_1 receives a pulse and input I_2 does *not* receive a pulse, the neuron will produce an output pulse *one time-step later.* This delay of one time-step is McCulloch and Pitts's formal counterpart of the processing time discussed earlier. The table also shows that for any other arrangement of pulses on the inputs, there will be no output pulse. In words: this neuron produces a pulse one time-step later *if and only if* there is a pulse on the first input and *no* pulse on the second input

We now add a second neuron that is the mirror image of the first. It produces a pulse *if and only if* there is a pulse on the second input and *no* pulse on the first input. We complete the circuit with a third neuron. This third neuron has fixed threshold 1 and two inputs, both of which carry weight +1 (Figure 5.5 c). It produces an output pulse if either or both of its inputs carry a pulse; it acts as an *inclusive or.*

We form the *exclusive or* circuit by connecting the outputs of the first two neurons to different inputs of the third neuron. We also "tie" input I_1 of the first neuron to input I_1 of the second neuron so that they are fed from one source; we treat input I_2 similarly (see Figure 5.5c). The resulting net has two input sources and one output. Inputs I_1 and I_2 are said to be *free* because the pulses on them must be supplied from outside the network; the other inputs in the net are constrained to the values supplied by the outputs to which they are connected.

With these connections the third neuron produces the *or* of the two

tic) and the storage registers (memory) for programmed digital computers.

The ability of the *exclusive or* network to record the parity of the total number of pulses it has seen since it was started up—a mathemetician would call it counting modulo two—can be extended. By stringing together copies of this *exclusive or* circuit, we can build networks that count to higher powers of two. (The simplicity of

possibilities recorded by the first two neurons. Because the inputs of the first two neurons are tied together, at most one of the two can produce a pulse at any given time. Thus, as a quick check shows, there will be an output from the third neuron, after two time-steps of delay, *if and only if* there is a pulse on the first input and *no* pulse on the second output *or* vice versa (see Figure 5.5c). As required, then, the third neuron produces the *exclusive or* of the pulses on inputs I_1 and I_2.

Indefinite memory occurs when we create a cycle in this *exclusive or* network by connecting the output of the third neuron to the second input I_2, forming a cycle in the network. Input I_2 is then no longer free, because a pulse appears on I_2 only when the third neuron produces a pulse. The only time the "user" sets the value of I_2 is at startup time: we have to say what the net's condition is at the outset, supplying what a physicist would call a *boundary condition*. Typically, we treat the net as quiescent initially, with no pulses circulating at startup time. That is, we let $I_2 = 0$ at the outset. Ever after, the value of I_2 is determined by the output of the third neuron.

With input I_2 constrained, the only pulses that come from outside the net are those specified at each time for input I_1 (see Figure 5.6). If the neuron is quiescent at the outset, time $t = 1$, then it remains quiescent until a pulse is supplied on I_1. When that pulse appears, it will continue to circulate around the loop until a second pulse appears on I_1. We say the neuron is *set* by the first pulse and *reset* by the second. Note that the network stays set *until* a second pulse occurs. If that second pulse never appears, the network stays set indefinitely. That is, the network remembers the occurrence of the first pulse for an indefinite amount of time. If a third pulse appears on I_1 after a second pulse, the network will be set again, and so on as subsequent pulses arrive (see Figure 5.6).

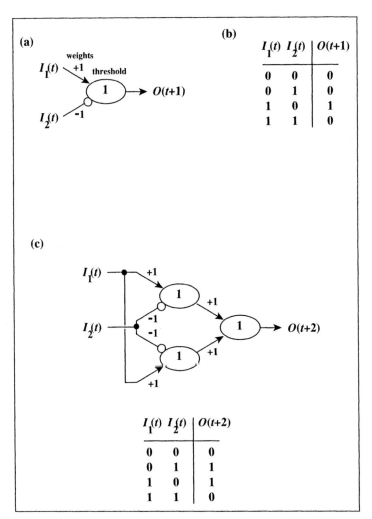

FIGURE 5.5
Network realization of "*exclusive or.*"

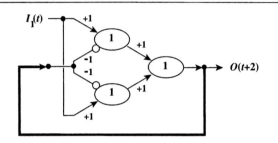

(The behavior of the net is easiest to understand if the pulses on $I_1(t)$ are constrained to occur only on even-numbered time-steps.)

When this net receives a pulse, it marks that occurrence for an indefinite amount of time by continuing to send an output pulse thereafter.

t	1 2 3 4 5 6 7 8 9 10 11 ...
$I_1(t)$	0 0 0 1 0 0 0 0 0 0 0 ...
$O(t)$	0 0 0 0 0 1 1 1 1 1 1 ...

If more than one input pulse is received, the output will only be "on" (1) if the total number of pulses is odd.

t	1 2 3 4 5 6 7 8 9 10 11 12 13 14 15 16 ...
$I_1(t)$	0 0 0 1 0 0 0 1 0 0 0 0 1 0 0 0 ...
$O(t)$	0 0 0 0 0 1 1 1 1 0 0 0 0 0 1 1 ...

FIGURE 5.6
Indefinite memory.

such a network is the primary reason general-purpose computers work with binary numbers.) With a few modifications to the circuit one can build a network that does arithmetic. It is in this sense that the *exclusive or* serves as a basic building block for programmable computers, providing both memory and arithmetic.

Following the line just laid out, we can build a network with cycles that can execute any program that can be written for a general-purpose programmable computer. Such a network can model any process that can be described by a computer-based model. In contrast, it can be proved (though I will not do it here) that there is no network without cycles that can exhibit indefinite memory; a fortiori, there is no network without cycles that can carry out the arbitrary computations of a general-purpose programmable computer. Indeed, a network without cycles (a feedforward net) provides only a minuscule subset of the capabilities that can be provided by a network with cycles.

Looking at Triangles—An Example

Now we are ready to look at emergent phenomena in a model neural network that mimics the central nervous system by having cycles. There is in cyclic nets a simple model of pattern recognition that demonstrates the critical roles of variable threshold, fatigue, and the extended Hebb's rule. This model relies on circulation of pulses—*reverberation*—in a cyclic neural net. It also illustrates the discovery of new features for describing the environment—the problem of learning new features that Samuel set aside.

Organization of the Model

Input. We'll assume that the input comes from an "eye" with a "retina" that consists of a large number of input neurons (see Figure 5.7). The input neurons will be arrayed to form a central area of sharp resolution surrounded by an area of much lower resolution, as in the mammalian eye. In a feedforward net this set of neurons would be called the input layer.

In this model, the eye will be used to observe line drawings of simple geometric figures, such as triangles and squares. It will be

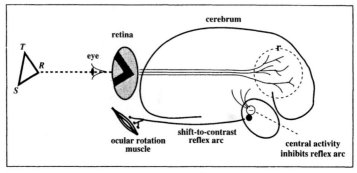

FIGURE 5.7
Behavior of a network with cycles (highly stylized).

sensitive to contrast in these figures, with lines offering higher contrast than the background, and the points where lines meet (vertices) offering highest contrast.

Processing. For simplicity, the model will do its central processing via a large number of neurons that are interconnected at random (see Figure 5.7). That is, the axon of any given neuron in the central group branches in such a way as to make random synaptic contacts with other neurons in the central group. The connections, once established, remain fixed thereafter. This arrangement, of course, sets aside all the structural biases that evolution has supplied to real nervous systems. There is, however, an advantage to using a model net that has no initial structural bias. Any feature extraction accomplished by such a net can only be the result of learning; it cannot be built in by an initial structural bias.

Note that random interconnection supplies the model net with a great many cycles of varying lengths. This, in fact, *does* mimic the highly interconnected cyclic nature of the mammalian nervous system (where structural biases are superimposed on the complex cycles).

Output. One further addition completes the overall organization of the model. We supply a reflex that controls the movement of the eye. It is similar to the human reflex that causes us to turn our vision toward any rapidly moving object in peripheral vision, as when we see a thrown ball or a flying bird "out of the corner of the eye." In the model I'll call this the *shift-to-contrast reflex*. This reflex, when activated, causes the eye to shift to a new point of contrast, say from one vertex in a line figure to another vertex in that figure.

The shift-to-contrast reflex is designed so that it is suppressed when neurons in the central processing part of the model are firing at a high rate. When the firing rate drops off, say through fatigue, the reflex is released. We can, somewhat fancifully, think that the reflex is released when central processing is tired of (bored with) the current observation. For simple geometric figures, the net effect of the reflex is to cause the eye to shift to a new vertex of the observed figure.

Emergent Behavior

Now we're ready to examine the effect of the large number of cycles in the central processor. We'll see three emergent phenomena: synchrony (groups of neurons entrain themselves into synchronous firing); anticipation (groups of neurons "prepare" to respond to an expected future stimulus); and hierarchy (new groups of neurons form to respond to groups already formed). We'll examine each of these phenomena as direct consequences of the organization of our model network.

The scenario begins when a geometric figure is presented to the eye. Because of the shift-to-contrast reflex, the eye will center on one of the vertices. Call it R. The light rays emanating from R will illuminate a particular contiguous set of neurons in the "retina" (the region of sharp resolution in the input layer—see Figure 5.8). As a result, these neurons begin firing at a high rate. The outgoing pulses are distributed, via their branching axons, to a substantial, randomly selected subset of neurons in the central processor. Call this subset r. The neurons in r also begin firing at a high rate, because of the large number of pulses arriving over the synapses formed by the input axons.

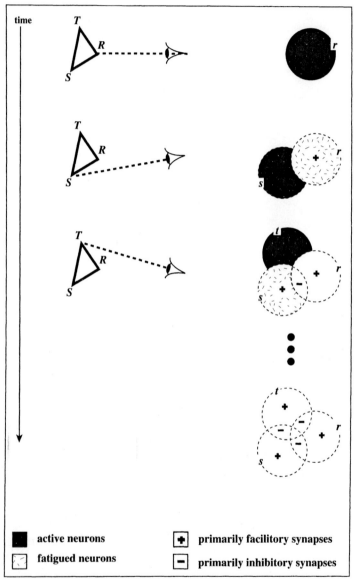

FIGURE 5.8
Input-controlled reverberation.

Synchrony. The neurons in *r* are interconnected with each other in loops, because of the pattern of random interconnections designed into the central processor (CP). So the neurons in *r* receive pulses from other neurons in *r,* in addition to receiving pulses from the input layer. Pulses in *r* recirculate through the loops of connections so formed. Observation shows that the combination of variable threshold and high firing rate causes some of these active CP neurons to begin to fire in a synchronized lockstep; they begin to reverberate. Neurons not participating in this lockstep group tend to be in the very high threshold part of the variable threshold when pulses arrive, so they do not fire. The result is a "weeding-out" process that favors neurons that belong to the lockstep group.

Because the neurons in the synchronized subset help fire each other, Hebb's rule causes these interconnections to be strengthened. The synchrony becomes self-reinforcing. As a result, subsequent presentations of a vertex in the same orientation will be more effective at causing this smaller, self-reinforcing subset to reverberate. The CP has learned to respond to the vertex by causing this subset to reverberate. In Hebb's 1949 theory, such a strongly interacting set of neurons is called a *cell assembly.*

If it were not for fatigue, this synchronous subset would go on reverberating indefinitely. However, the high firing rate causes fatigue gradually to increase, progressively raising the thresholds of the neurons involved. The fatigue increases until the firing rate of the component neurons drops to a nominal level. Synchrony fails at that point and the whole assembly ceases to reverberate.

A new vertex. Now the shift-to-contrast reflex comes into play, because the reflex is no longer inhibited by high activity in the CP. The eye shifts to a new vertex—call it *S*—projecting a new pattern on the retina's area of sharp resolution (see Figure 5.8). This, in turn, causes a different subset of neurons in the retina to fire at a high rate. The axons of these input neurons project to a different subset of neurons in the CP. Call this subset *s.*

Reverberation is established in *s*, much as it was earlier in *r*, with one important difference. Pulses going over connections from group *s* to group *r* will not succeed in firing neurons in *r* because they are highly fatigued. The downside of Hebb's law now comes into play. These particular synapses decrease in strength; after several presentations of the pattern they can actually take negative values. That is, when group *s* is active, it comes to actively inhibit the firing of neurons belonging to group *r*.

The same effect takes place when the shift-to-contrast reflex centers on the vertices in reverse order, with *S* preceding *R*. A minute's thought shows that, under these circumstances, a given CP neuron can belong to group *s* or group *r*, but it cannot belong to both. We encounter the "*exclusive or*" again. Hebb's law, in combination with synchronous firing and fatigue, causes groups *r* and *s* to have no overlap. The CP learns a distinct, disjoint representation for each vertex. Moreover, these representations come to cross-inhibit each other, because the synapses between them take on a negative value. Only one of the groups can reverberate at a time.

If the figure is a triangle, this scenario can be extended to all three vertices *R*, *S*, and *T*. As the shift-to-contrast reflexes scan back and forth over these vertices, the CP develops three groups *r*, *s*, and *t* that are reverberatory and mutually exclusive (see Figure 5.8).

Anticipation. The combination of fatigue and mutual inhibition between the groups *r*, *s*, *t* has an unexpected consequence: anticipation. Anticipation is the counterpart of lookahead in Samuel's checkersplayer—an expectation about future actions. In both cases we have virtual projection into the future that lets the system avoid the consequences of irreversible mistakes ("falling off a cliff"). The example here is too simple to show such large objectives, but it makes the point.

Anticipation begins to show up after some group, say *r*, has been strongly reverberating for a time. While *r* is reverberating, it actively prevents groups *s* and *t* from reverberating. Indeed, because *s* and *t* are being actively inhibited by the pulses from *r*, neurons in these groups are firing at a substantially *lower* rate than other neurons in the CP. As a consequence, their fatigue drops to an abnormally low level compared to the other neurons.

When *r* begins to fatigue, the active inhibition of neurons in *s* and t begins to decrease. The abnormally low fatigue of these neurons, and the resulting lower thresholds, make them much more sensitive to incoming pulses than the other neurons in the CP. Even a weak stimulus from the eye, in one of the input regions corresponding to S or T, will start neurons in one of these groups firing, in preference to other sets of neurons in the CP. The CP, via these hypersensitive groups, anticipates that the next stimulus will come from one of the other vertices.

If the input neurons now favor s, the firing rate in s picks up just as the inhibition from r decreases. The increased firing in s further dampens the firing rates in r because of the inhibitory synapses on the connections from s to r. The firing rates in r are depressed still further, and so it goes, enforcing a rapid transition from activity in r to activity in s. Transition is complete.

Because the same story holds for s and t in turn, the result is a sequence of three reverberating groups, a sequence that is synchronized with the scan of the three vertices. Because small differences can cause the shift-to-contrast reflex to visit the vertices in different orders, the sequences will be mixed, looking something like $r,s,t,s,t,r,s,t,\ldots$. Still this set of three reverberating, mutually exclusive cell assemblies embodies the three-ness of the triangle.

More important, if r is firing, it prepares the CP for the firing of s or t by actively depressing the accumulation of fatigue in s and t. In effect, when the CP "sees" one vertex of the triangle, it anticipates that other vertices will follow. It will take a strong stimulus from some other part of the environment to break up the effects of this anticipation prematurely. It may even be that the CP will compensate for a vertex that produces a substandard stimulus. That is, if S is the substandard vertex, the group s will still reverberate with near-normal intensity. This "filling in" is reminiscent of a familiar phenomenon in the psychology of human vision, where the observer "sees" a vertex that is faint or missing.

Hierarchy. Once the r,s,t sequence is well established through the action of Hebb's rule, a further step becomes possible, a step that depends on a rather special group of neurons. There are individual neurons in the CP that receive connections from all three groups r, s, and t (see Figure 5.9), as a consequence of the CP's

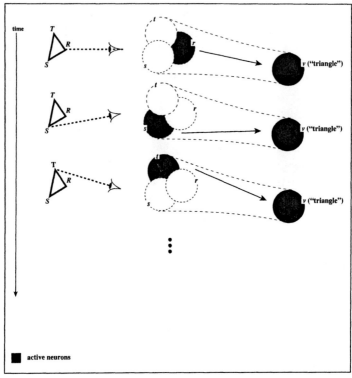

FIGURE 5.9
Hierarchical reponse to cell assemblies.

random interconnection scheme. Moreover, some of these neurons have axons with few or no branches leading back to any of the neurons in r, s, and t. We use v to designate the set of such neurons that have no back-connections. Although the interconnection pattern of neurons in v is not typical, there will actually be a fair number of such neurons if the CP involves, say, 10,000 or more neurons. The neurons in v are atypical, but not so much so that they are statistically unlikely.

Let's look at the firing pattern of neurons in v. These neurons are stimulated by the neurons in r, s, and t, but they do not have the reciprocal connections that allow them to participate in the reverberations. They do fire at an above-average rate when neurons in *r*, *s*, or *t* are reverberating, but they do not fire at the intense rate enforced by the lockstep. As a result, the neurons in *v* accumulate fatigue much more slowly than the reverberating neurons. This, in turn, allows neurons in *v* to fire at an above-average rate for a substantially longer period, as long as they are being stimulated by neurons in the reverberating groups. The above-average firing in *v* can persist through a period that overlaps several steps of reverberation.

The regularity of the *r,s,t* sequence, once under way (enforced by the learned cross-inhibition), provides sustained input to the neurons in *v*. Hebb's rule strengthens the connections between the reverberatory groups and the neurons in *v*, further reinforcing the effect. Because the neurons in *v* continue to fire at an above-average rate as the sequence advances, these neurons respond to the whole three-ness pattern, rather than to the individual vertices.

Said another way, the neurons in *r*, *s*, and *t* translate the pattern provided by the shift-to-contrast reflex into something to which the neurons in *v* can respond as a whole. The CP has built a hierarchy, where the neurons in *v* abstract from, and "stand above," the details provided by the eye.

Summing Up

There is much more to this story, but we've gone far enough to touch on some of the main points that bear on emergence.

1. By providing the network with a large number of cycles (loops) in its interconnections, we open the possibility of reverberation in subsets of neurons. This kind of activity provides for indefinite memory, and it makes possible the organization of neurons into the cooperative assemblies that serve as building blocks for organized sequential behavior (see Hebb, 1949).

2. Three simple mechanisms provide an astonishing increase in the range of behaviors a neural network can produce.

a. *Variable threshold.* A neuron's threshold decreases as the time since it last fired increases. This decreasing threshold makes the neuron increasingly sensitive to incoming pulses when it remains quiescent over an extended period. The variable threshold allows the neuron to act as a frequency modulator, firing at a rate that reflects the average synapse-weighted strength of the pulses impinging on its surface.

b. *Fatigue.* A neuron that fires at a high rate *over an extended period* has its threshold steadily incremented—in effect the whole variable threshold curve is translated upward. Contrariwise, a neuron that fires at a low rate over an extended period has its threshold steadily decremented. Fatigue eventually forces a neuron's firing rate back to a "normal" or "set-point" level: no neuron can continue to fire at a rate above or below this set-point.

c. *Hebb's rule.* If neuron X fires at time t and neuron Y fires at time $t+1$, then any synapses that X's axon makes at Y are strengthened. Contrariwise, if X fires at time t and Y does *not* fire at time $t+1$, the same synapses are weakened. Hebb's rule acts without knowledge of any firing beyond the two neurons directly involved. Using Samuel's notion of credit assignment, the rule assigns credit (synaptic weights) without requiring any larger knowledge of the system's activities; it acts without the direct intervention of a referee or executive.

Fatigue in combination with Hebb's rule favors and reinforces reverberating assemblies of neurons. By appropriately sequencing external stimuli, these assemblies can be conditioned into sequences. The sequencing of stimuli may occur automatically, as in the case of the shift-to-contrast reflex and the natural jump-focus-jump (saccadic) movements of the eye. Fatigue and Hebb's rule also lead to cross-inhibition between assemblies stimulated by mutually exclusive objects (as in the case of the vertices of a triangle). As cross-inhibition develops, it forces inactive assemblies in a sequence to fire at abnormally low rates. In consequence, these assemblies acquire abnormally low fatigue, becoming more ready to reverberate than assemblies not in the sequence. Here we see the beginnings of anticipation: elements in the stimulus sequence not yet seen are anticipated by the corresponding extrasensitive assemblies.

3. Well-developed sequences of reverberating neuron assemblies can act to stimulate other neurons in regular ways. Hebb's rule then acts to generate new assemblies that respond to the sequence as a whole. For example, a triangle, through the mediation of saccades and a shift-to-contrast reflex, produces a sequence of three reverberating cross-inhibitory assemblies. This sequence, once developed, produces another assembly that responds to the sequence as a whole, an assembly that responds to the three-ness of the triangle. Here we have the beginnings of a hierarchy, generated in response to regularities in the external stimuli.

4. Situations not previously encountered can, despite their novelty, cause well-developed assemblies to be sequenced or pieced together in new ways. The assemblies become building blocks, responding to familiar elements in new situations. This process is a precursor of that everyday, but astonishing, human ability noted in the *Summing Up* section at the end of Chapter 4: humans effortlessly parse unfamiliar scenes into familiar objects, an accomplishment that so far eludes even the most sophisticated computer programs.

The behaviors possible for neural networks with cycles are subtle consequences of the interaction of prosaic mechanisms. The fact that cycles make feedback possible does little to prepare us for the phenomenon of indefinite memory—the ability to retain memories of past events for indefinite periods. Yet counting, arithmetic, general-purpose computing, and ultimately mental models that permit virtual exploration of the environment are impossible without indefinite memory. Likewise, the common notion that fatigue decreases performance does little to prepare us for its role in generating anticipation.

Comparisons

Neural networks so constructed give us a second in-depth example in which "more comes out than was put in." When Samuel's checkersplaying program began to beat him consistently, it clearly exceeded any expertise he could have supplied directly. Similarly, when a neural network begins to form stimulus-related reverberating assemblies, it acquires behaviors and organization not present in the initially supplied random connections. In both cases we see simple mechanisms generating behaviors that transcend the capacities of the designer.

The complexity that results from the interactions of these simple mechanisms is neither more nor less mysterious than the complexity that emerges when a game is defined via a few rules. In games the possibilities are completely determined by the rules, "put in" at the start. However, it takes intensive study, and time, to reveal these possibilities. As we uncover these possibilities, we see unanticipated regularities and symmetries. The study of chess over the centuries has revealed patterns that contribute to winning: linked pawn formations, sacrificial gambits, means of enhancing mobility, and so on. Similar regularities and patterns emerge when we study models based on interacting mechanisms. The laws embodied in Newton's equations, though intensively studied since their conception, still reveal new regularities and possibilities.

We gain further insight into these commonalities if we return to the descriptive technique we used in discussing games, a description in terms of *state* and *strategy*. To review: The rules of the game determine the arena—the *environment*—in which the checkersplayer must operate. The configuration of pieces on the board at any given time determines the *state of the environment* at that time. The checkersplaying program learns of the state of environment via a set of feature detectors, and a weighted sum of the feature-detector readings determines the checkersplayer's *strategy*. The strategy determines actions that change the state of the environment (the arrangement of pieces).

How do we describe the neural net in similar terms? To begin, let's treat the central processor as the counterpart of the checkersplayer. Then, in our example, the CP's environment is the figure presented, and the state of the environment, from the CP's point of view, is the particular part of the figure (vertex) being viewed. The onset and offset of reverberation in different learned neuron assemblies determines changes of the point of focus in the figure, so these assemblies determine the CP's strategy. As in Samuel's checkersplayer, the strategy determines changes in the state of the environment.

This comparison is more than an exercise in terminology. The concepts of state and strategy are well-established technical concepts with a considerable lore, and they are broadly applicable. Most of all, these concepts help us separate the details that are idiosyncratic to particular systems from the parts the systems hold in common. This is an essential step in determining the common parts that contribute to emergence.

Onward

Emergence must somehow be bound up in the selection of the rules (mechanisms) that specify the model, be it game or physical science. Here we come upon our central mystery in a different guise. How do humans, with their limited capacity for tracing out complexities, manage to select interesting rules and mechanisms? How does the scientist select the "laws" that are so effective at uncovering unexpected phenomena in the world?

The whole process of constructing a rule-based model starts when our attention is attracted to some complex pattern in the world—the movement of armies in battle, or the patterns of the "wanderers" (planets) in the night sky. At this point we do not have the benefit of the hindsight provided by years of study of the candidate rules. We could attempt to derive some of the deeper consequences of different rule choices before making a selection, but to do so would consume inordinate amounts of time. The actual selection process is totally different. Derivation and deduction have, at best, a limited role; at the time of selection, we have only a sketchy idea of the possibilities that will emerge. Some other process must mediate the transition between the patterns of interest and the rules that attempt to model those patterns.

Almost by definition, the selection process must be a matter of induction—moving from particulars to abridged, more abstract descriptions. Once again, we encounter the importance of dropping selected details. Knowing what details to ignore is *not* a matter of derivation or deduction; it *is* a matter of experience and discipline, as in any artistic or creative endeavor. When this process goes well, the resulting description reveals repeated elements and symmetries that suggest rules or mechanisms, a matter to which we will return.

With the examples of the checkersplayer and neural networks as prologue, let us now place emergence in a broader setting. Then can we distinguish the particularities of the examples from the elements that are essential to emergence. The broader setting will help us make connections between the model-building process and those other activities where we sense emergence, such as the creation of metaphor and the construction of stories and poems.

CHAPTER 6
Toward a General Setting

Lᴇᴛ's ʀᴇᴠɪᴇᴡ some of the common elements of the examples of emergence we've so far examined:

- Each example, in one way or another, models the world. Even a game like chess has its origin in the powers maneuvering on early battlefields, and its later European incarnation retained warlike names for its pieces: knight, castle, pawn (foot soldier), bishop (in the Crusades), and so on.
- Each model consists of multiple interacting copies of a limited number of pieces, particles, or components (types). In board games such as chess and checkers, we have the designated game pieces. In neural networks, we have the prototype neuron with specified properties (variable threshold, fatigue, and Hebb's rule in the example of the previous chapter—sophisticated networks may involve several types of neurons). In the models of physics and chemistry, we have elementary particles and atoms. The complexities of the model are generated by the interactions of these copies.
- The configuration of the model's components changes as time elapses. With careful modeling, the particular configuration of the components at any time fully determines what can happen next. In a board game, we need only know the arrangement of the pieces to know what legal moves we can take from there on. In a neural network model, the firing state of each neuron (pulse or no pulse), along with its threshold, fatigue, and synaptic weights, determines what will happen next. In classical physics and chemistry, we need only know the posi-

tion and energy of the particles involved. That is, the *state* of the model is given by the configuration, and future possibilities depend only on that state, not on how it was achieved.

■ Interactions are constrained by a succinct list of rules (equations). All possible state (configuration) sequences are the outcomes of a succession of limited rearrangements specified by these rules. In board games, these are the rules that define the game. In the checkersplaying program, an additional set of rules (computer subroutines) determines changes in the feature weights, using a lookahead tree based on the rules of the game. In neural nets, we have the rules that determine when any given neuron fires, along with an additional set that determines how the synaptic weights change. We have borrowed a word from game theory, *strategy,* to describe these additional rules in the case of checkers and neural nets. Later we will relate all of this to the more technical notion of a *transition function.*

Prior to the advent of programmed computers in the 1940s, models were severely limited in the number of component types and governing rules. Sometimes a model was carefully contrived to allow mental exploration of theories, as in Einstein's famed thought (gedanken) experiments on quantum theory (see Jammer, 1974). At other times, the model was designed for mathematical (pencil-and-paper) exploration of theories and laws, occasionally backed by extensive hand calculation. But the human hand, driven by the human brain, can do only so much in an hour. Even limited numbers of components and rules generate more possibilities than can be revealed by a lifetime of pencil-and-paper work. Thus, the dozen or so rules of games like chess and Go, or the five axioms of Euclidean geometry, continue to provide surprises after centuries of study.

Agent-Based Models

The limitations of a pencil-and-paper approach become particularly evident when the model's components are multitudinous and mobile. Think of a market. The negotiations that determine the state of the market depend on the objectives and strategies of individual agents—the buyers and sellers—as they enter and leave. Large markets, like the stock and commodity markets, are notorious for their complexity and unpredictability. Even simple models of a market, with buyers and sellers severely limited in their actions, exhibit complicated dynamics (see Arthur et al., 1997). There are sudden shifts in activity and price levels, with occasional crashes and bubbles. The model market may go on for a thousand simulated days with only minor fluctuations, and then shift suddenly into a new regime, much like a real market. All of this is well beyond the most dedicated pencil-and-paper explorations, be they calculations or mathematical analysis.

Still, much that goes on in this world of ours is naturally described by the interaction of individuals with individual strategies, as in games. When we model such systems, these individuals go by the name of *agents,* and the models are called *agent-based* models. The agents that make plans in social organizations (governments, businesses, and the like) are the people. At times, though, we shear away some details and speak of the plans of a department or even a whole government (in the case of international relations). At each level we can design interesting agent-based models. The same can be said for ecosystems, where we might select interacting species as agents, or the immune system, where the antibodies act as agents. We even gain some insights when we treat the interacting genes in a chromosome, or the organelles in a cell, as agents. And, of course, each of these systems exhibits emergent phenomena.

The scope of agent-based models, and their obvious relevance for the study of emergence, suggests we give them serious consideration when designing a general setting, if we can handle the complexity. The ant colony discussed at the beginning of the last chapter illustrates the point. The rules describing an individual ant's repertoire can be few; the complexity of the colony emerges from the large numbers of ants, and their coupled interactions with each other and the environment. In this the ant colony has much in common with a neural network, where the flexible behavior of the whole depends on the activity of large numbers of agents (neurons in the case of the neural network) described by a relatively small number of rules. A substantial part of the complexity in most agent-based models stems from the sheer number of agents involved.

When there are large numbers of agents, simple or not, the "move tree" (the range of possible interactions) far exceeds the already enormous move trees associated with checkers or chess. Because the actions of the individual agents are conditioned by the immediate surroundings (other agents and objects in the environment), there is no easy way to predict the overall behavior by looking at the behavior of an "average" individual. The difficulty increases enormously when individual agents can learn or adapt. Then an agent's strategy is not only conditioned by the current situation, it can also change over time, as in the cases of the checkersplayer and the neural networks. As the difficulties increase, so do the possibilities for emergent behavior.

There are real incentives for studying agent-based models, though it is close to impossible to tease out the complexities of such models using pencil and paper alone. The advent of the computer changes all that. We now have a tool that makes possible detailed studies of agent-based models.

Enter the Computer

The forte of a programmed computer is repetitive action. When we write programs, we write many subroutines—sequences of instructions that are executed again and again until some condition is met. Such a subroutine may put an array of bright dots on a screen, forming the letter *A*, or it may calculate the fatigue of a modeled neuron. We combine these subroutines to attain the final product, be it a screen-oriented word processor or a neural network simulation.

In principle, we could use pencil and paper to carry out the instructions of any such program. It is, after all, only a matter of following instructions. However, the process would be time-consuming beyond feasible human persistence or devotion. Certain individuals in the previous century devoted years, even decades, to calculating additional digits in the number π, a number that is complex because it cannot be presented as a simple ratio of integers. Though this work did reveal some interesting properties of π, it was not a creative task in the usual sense; it was a matter of executing a simple routine over and over again.

The same result using the same routine can be accomplished with a modern laptop computer in a matter of minutes. Pocket calculators often use subroutines to calculate instantaneously eight or ten digits of numbers like π, instead of using storage registers to hold the numbers explicitly. Indeed, pocket calculators can use programs to calculate, on the spot, the numbers that were recorded in the extensive tables (for instance, Burington's 1946 tables) that engineers used to carry to augment their slide rules. Both the tables and the slide rules have been replaced by pocket calculators.

Such speed offers a qualitative change in what is possible. We can carry out investigations where routines requiring many repetitions, like the routine for calculating π, are a small part of a larger effort. In particular, we can build models in which hundreds of subroutines are tied together. Such programs have a modular nature that suits agent-based modeling. First, we translate the rules defining individual agents into subroutines (sequences of instructions somewhat like the subroutines that defined features for Samuel's checkersplayer). Then we tie these subroutines together with additional instructions (see Holland, 1995) to provide for the interaction of the agents. If the computer has many processors—if it is a *parallel computer*—we may even allocate one processor to each agent, shuffling information back and forth between the processors as the agents interact. Whatever the hardware organization of the computer, its tremendous speed allows the exploration of models with large numbers of agents.

For such explorations, computer-based models provide a halfway house between theory and experiment. Computer-based models are not experiments in the usual sense because they do not directly manipulate the world being modeled. Nevertheless, as the model is executed, patterns and symmetries will typically show up in the ongoing action (like the synchrony and reverberation that emerge in neural networks with cycles). Such emergent characteristics can suggest experimental designs for real situations. Though the computer-based model is not a physical experiment, it is completely rigorous; there is no ambiguity about the relation between the definition of the model and the results it implies. Run the same model with the same settings twice and you will get the same results.

In its rigor, a computer-based model is similar to a mathematical (equation-based or axiomatic) model, but its results hold only for the particular settings used when it is executed. A mathematical model, on the other hand, yields results that hold over some well-specified domain. Still, by executing the computer-based model several times, with different initial settings, we may discern patterns and regularities that recur in the results. For instance, we may see that Samuel's checkersplayer consistently assigns high weights to certain kinds of features, or we may see that a neural network with loops consistently develops small, reverberating assemblies. These regularities can provide clues for constructing a mathematical model in which the regularities can be *deduced* from the structure. That is, the regularities suggest theorems that can be deduced from the axioms (rules) defining the model.

Because the structure of a computer-based model consists of rigorously specified modules (subroutines), we gain certain advantages. We can dissect the model to search for the source of observed regularities. We can test hypothesized causal relations by varying the structure and parameter settings of selected subroutines. If the variants produce consistent changes in the observed regularities, we can incorporate the variants in a more general mathematical model; often the selected subroutines can be translated into functions that capture the observed relations. With insight, and luck, we can build a mathematical theory around these functions.

Computer-based models greatly expand our opportunities for using intuition. They fill a role similar to the role earlier played by gedanken experiments and "back-of-the-envelope" calculations. But the speed of the computer vastly expands that role. The computer-based models can include details, interactions, and numbers of agents that would simply overwhelm a pencil-and-paper version. There is a danger in this, the danger of including large numbers of details simply because it is feasible to carry out the corresponding calculations. The inclusion of too many details can defeat the whole purpose of building a model, a danger we'll discuss later in Chapter 10. But prudence can allay the danger, and the danger does not foreclose the opportunities.

There is an added advantage. The unforgiving rigor of a computer program forces careful thought. No amount of clever rhetoric or wishful thinking will cause a computer model to deviate one iota from the consequences of the rules it embodies. Though the computer-based model does not offer the general conclusions of a mathematical model, it does enforce a similar discipline.

Emergence and Nonlinearity

Before the advent of programmed computers, we could only study models with large numbers of agents by assuming that the individual agents exhibited a typical or "average" behavior. The overall behavior was seen as the sum of these average behaviors. An analysis of the summed behaviors often provides useful information about multiagent systems; the methods of statistical mechanics in physics (see Feynman et al., 1964) and the use of matrix methods to study ecosystem interactions (see May, 1973) provide convincing examples. However, as I have pointed out, this averaging has limitations. The behavior of an ant colony is not the simple sum of the behaviors of a group of average ants. The coupled interactions of the ants provide a coherence to the nest that far exceeds anything predictable in terms of simple summations.

Emergence is above all a product of coupled, context-dependent interactions. Technically these interactions, and the resulting system, are *nonlinear*: The behavior of the overall system *cannot* be obtained by *summing* the behaviors of its constituent parts. We can no more truly understand strategies in a board game by compiling statistics of the movements of its pieces than we can understand the behavior of an ant colony in terms of averages. Under these conditions, the whole is indeed more than the sum of its parts. However, we *can* reduce the behavior of the whole to the lawful behavior of its parts, *if* we take the nonlinear interactions into account.

Requirements for a General Setting

With this review of salient points in mind, what can we say about a general setting for the discussion of emergence? What form should it take? To answer these questions, I'll first emphasize some general requirements, then list some specifics. In the next chapter, the specifics will be translated into components of the setting I want to propose.

By now it's clear that the setting must take modeling as its central concern: the setting must provide a way to model a wide range of real systems that exhibit emergence. Models as different as Samuel's checkersplayer and neural nets with cycles must fit comfortably within the framework. Because systems that can be modeled in terms of agents frequently exhibit emergent phenomena, it is also clear that the setting must handle agent-based models in a simple, direct way. Finally, the setting must capture the organized perpetual novelty that we expect from games and other rule-based models.

Beyond these generalities, one recurring theme is essential to emergence: in each case there is a procedure for freely generating possibilities, coupled to a set of constraints that limit those possibilities. In checkers, we have the set of possible arrangements of pieces on the board constrained by the rules of the game. In neural networks, we have the possible ranges of behavior of individual neurons (firing rates) constrained by their connections to other neurons. The general setting must be built around this theme.

The examples and discussion so far suggest the following specifics.

- *State.* When we looked at board games, there was a direct translation from the arrangement of pieces on the board to the state of the game. The state of a system, when properly defined (whether for games, neural nets, or physics), subsumes all aspects of past history that are relevant to future possibilities. This feature greatly simplifies the study. A general setting for the study of emergence should make the definition of state as straightforward as it is for games.
- *Game tree (transition function).* The game tree unifies the study of a vast range of games, games as apparently distinct as chess and poker. It makes possible a careful definition of strategy, providing the foundation for the minimax theorem (von Neumann and Morgenstern, 1947); it is also a major part of the conceptual apparatus Samuel used in developing his lookahead procedure. The general setting should provide similar apparatus for studying other systems that exhibit emergence. We'll soon see that the appropriate generalization is the notion of a *transition function.*
- *Rules (generators).* The game tree is specified implicitly via a small set of rules. It is the thesis of this book that the study of emergence is closely tied to this ability to specify a large, complicated domain via a small set of "laws." In addition to games and neural nets, we've already mentioned Euclid's axioms and the equations of Newton and Maxwell, as models governed by simple rules. Such models, though simply defined, are complex enough to provide new insights, even after long periods of intense study. In the next chapter, I'll use the notion of a *generating procedure* to capture the use of laws to generate complex domains.
- *Agents.* Ants in a colony, neurons in neural networks, or particles in physics—all are described in terms of rules or laws that

determine their behavior in a larger context. In each case we can describe these agents as processing material, energy, or information to produce some action which is usually the transport of material, energy, or information. More generally, we can describe the agent as processing an input to produce an output. We can go further using the notion of state: we can talk of an *input state* being processed to produce an *output state.* The input state is determined by the immediate environment of the agent, and the output state determines the agent's effect on its immediate environment. When an ant detects food (the input state), it begins to lay a scent trail back to the nest (an output state affecting the environment). The interactions of a given agent are then described in terms of the effects of other agents on its input state, in just the way we described the effect of pulses arriving at the surface of a neuron.

With these observations as prologue, we are ready to start laying out a general setting for the study of emergence via *constrained generating procedures.*

CHAPTER 7

Constrained Generating Procedures

THE GREEKS, as mentioned in Chapter 1, already knew the benefits of describing diverse objects (machines) in terms of elementary mechanisms (the lever, the screw, and so on). This chapter develops a similar setting for describing the diverse systems that exhibit emergence. The Greek approach to machines also provides a helpful, intuitive guideline for developing this setting. It suggests looking at emergence in terms of *mechanisms* and procedures for *combining* them. To make this work, we have to extend the idea of "mechanism." The generalization used here comes close to the physicist's notion of an elementary particle (say a photon) as a mechanism for mediating interactions (causing an electron to change its orbit around an atom). We can use this extended notion of mechanism to provide precise descriptions of the elements, rules, and interactions (for example, agents) that generate emergent phenomena. And we can compare quite different systems, increasing our chances of finding common rules or laws that describe systems exhibiting emergence.

The only feasible way to attain the requisite generality and precision for a general setting is to adopt a suitable mathematical notation. Though this chapter contains a hefty dose of such notation, it does *not* require knowledge of anything more than elementary mathematical concepts, such as function and set, concepts we've already encountered. In particular, this chapter does not require knowledge, or recall, of mathematical theorems or sophisticated mathematical operations. If you have not thought about things like functions and sets since high school, this section *will* require some effort to match your intuition with the notation, but in all other respects it is self-contained.

The observations of the previous chapter provide detailed guidelines for this effort. I'll translate those requirements, step by step, into mathematical counterparts, spending some time to show that the translation captures the relevant ideas. We'll be mimicking the path that scientists often follow in moving from intuition to precision. The result is an explicit definition of a broad class of models I'll call *constrained generating procedures* (*cgp*'s). The models that result are dynamic, hence *procedures*; the mechanisms that underpin the model *generate* the dynamic behavior; and the allowed interactions between the mechanisms *constrain* the possibilities, in the way that the rules of a game constrain the possible board configurations. All the systems we've looked at so far can be described as *cgp*'s. Potentially, *any* constrained generating procedure can exhibit emergent properties.

In outline, the path to constrained generating procedures involves the following steps.

1. First, we'll translate the notion of rule (for instance, the rule for jumping in checkers, or Hebb's rule) into the notion of *mechanism*. As with rules for games, or laws for a physical system, mechanisms will be used as the defining elements of the system.

   Roughly, a mechanism responds to actions (or information), processing that input to produce resultant actions (or information) as output. The simple lever, one of humankind's earliest discoveries, provides an easy example (see Figure 7.1): when you pull down on one end of the lever (the input), you get a force at the other end (the output) that is multiplied by the ratio of the two lever arms (the processing). More complicated mechanisms may have several inputs and may produce several different outputs—think of a mechanism that sorts coins. The word "mechanism" has variant meanings in common usage, but

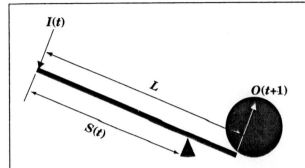

$I(t)$

L

$O(t+1)$

$S(t)$

The output $O(t+1)$ of the mechanism is determined by a transition function f that relates the position of the fulcrum $S(t)$ and input force $I(t)$.

$$O(t+1) = f(I(t), S(t)) = \frac{I(t) \cdot S(t)}{L - S(t)}$$

The transition function is derived from the "conservation of moments" law for levers, a step along the way to the powerful "conservation of momentum" law of modern physics.

FIGURE 7.1
A simple mechanism—the lever.

I'll use a definition and notation that restricts the word to a single, unambiguous meaning.

2. Next, we'll define ways of linking mechanisms together to form networks; these networks are the constrained generating procedures, the *cgp*'s. Because most models involve more than one kind of mechanism (as the different elementary particles in physics), the setting must make evident how the actions of one mechanism can affect others.

Within this setting, it is the interaction of the mechanisms that generates complex, organized behavior. Typically, the mechanisms allowed will be few in kind and simple to describe, enforcing the deletion of many details. Still, the interactions provide possibilities not easily anticipated by inspection of the individual mechanisms. The complexity increases when multiple copies of the basic mechanisms are allowed, as we've already seen in the case of ant colonies and neural networks.

We'll first look at *cgp*'s in which linkage captures the constraints imposed by the fixed landscape of a board game, or the geometry of a physical system. Later we'll look into a generalization whereby links can be made and broken from within the *cgp*, changing the underlying geometry. This facilitates models based on mobile agents.

3. Once the mechanisms are linked, we come to the counterpart of the game tree—the set of possibilities generated by the constrained interactions of the mechanisms. The examples already examined show that the tree of states is a handy way of modeling the possible courses of action—the strategies and dynamics—of games. Now we want to extend that notion to systems in general.

To accomplish this objective, we have to define the state of a whole *cgp* in terms of the states of the component mechanisms. We have to be able to do what we did for the checkersplayer and neural nets: condense everything about the *cgp* that is relevant to future possibilities into a single entity called its global state. Then we go on to describe the legal ways of making transitions from state to state. Following the suggestion at the end of the last chapter, we'll see that the possibilities for state changes are described precisely by a *transition function*.

4. Finally, we want to provide a specific procedure in *cgp*'s for defining hierarchies of subassemblies, using more complex mechanisms built up from the basic mechanisms. Advantages in explanation will accrue, along the lines noted by Simon in

1969, and the resulting organization parallels the hierarchical nature of most systems exhibiting emergence.

The major advantage here is again the one the Greeks perceived. It should be possible to treat a whole *cgp* as a mechanism that can be used to build still more complex *cgp*'s. This treatment parallels the use of subroutines, themselves composed of elementary instructions, as elements of a computer program. By making such definition a part of the *cgp* apparatus, we can capture the hierarchies that are major features of systems exhibiting emergence, like the molecule, organelle, cell, organ, organism, . . . hierarchy in biology.

Mechanisms

The term "mechanism," as used here, is defined by means of a *transition function*, which is similar in concept to the strategy function described in Chapter 3. In the case of a game, we had to define the state of the game before we could define the move tree and the strategy function. Similarly, for a mechanism, we must first define the *state of the mechanism* before we can define the transition function.

As a simple example of the state of a mechanism, consider the state of a watch. The watch has (a) a wound spring that transmits power (b) through a ratchet that advances (c) to produce the rotary motion of an indicator ("hand"). The state of the mechanism includes the tightness of the spring and the position of the ratchet, as well as the position of the indicator.

When we look at the watch as a mechanism, we see a succession of states as the hands move around the face of the watch. Internally, this movement of hands is caused by movement of the ratchets and gears, unwinding of the mainspring, and so on—all components of the state. This *state trajectory*, the sequence indicated by the movement of the hands, is the dynamic of this mechanism. The watch is somewhat atypical of mechanisms because its input is intermittent. The trajectory of states proceeds autonomously between windings of the mainspring, whereas mechanisms more typically are under the control of a steady stream of inputs (similar to the succession of decisions that determine the unfolding of a game). Still, the watch provides an example of the kind of dynamic we can expect from mechanisms, and illustrates the way in which simpler mechanisms (the mainspring, ratchets, and so forth) can be combined to yield a more complex mechanism. Much as the combination of rules defines a game, combinations of complex mechanisms built from simple ones have a major role in *cgp*'s.

The paragraphs that follow generalize and formalize these ideas, to provide a broad, precise definition of mechanism that underpins our general setting.

To develop the notation, we must represent the possible configurations as states of the mechanism. Assume the states can take on only a finite set of distinguishable configurations, a requirement that must be satisfied if *cgp*'s are to be modeled on a computer. That is, limit the amount of detail that can be captured in the model. Then we can represent the set of states S by an "alphabet" $\{s_1, s_2, s_3, \ldots\}$, where one letter (symbol with index) is assigned to each possible state.

The transition function, f, takes the current inputs to the mechanism, as well as the current state, as its arguments and produces the next state of the mechanism. So, to proceed with the definition, we must represent the possible input options to the mechanism as well as its possible states. We can proceed as for the states, allocating an alphabet of possibilities to each input. Thus, input j will have an associated alphabet $I_j = \{i_{j1}, i_{j2}, i_{j3}, \ldots\}$, where the j subscript to the letter i_{jh} indicates that letter as one of the possible values for input j; the second subscript, h, indexes those possibilities. We can think of these letters (symbols with double index) as naming possible *input states*. Thus, i_{j2} names state 2 of input j.

If the mechanism has k inputs, then there will be k alphabets $\{I_1, I_2, \ldots, I_k\}$, with one alphabet for each input. The alphabets may all be distinct, though that is not required. A cleaner notation results if we have a way to designate the set I of *all* possible *combinations* of input values that can be presented to the mechanism. Using a standard piece of mathematical notation, we define this set of combinations, I, by using the "product" of the input sets, $I_1 \times I_2 \times \ldots \times I_k$. For example, if $I_1 = \{a, b, c\}$ and $I_2 = \{x,y\}$, then

$$I = I_1 \times I_2 = \{(a,x), (a,y), (b,x), (b,y), (c,x), (c,y)\},$$

the set of all pairs that can be formed by first picking an element of I_1 and then picking an element of I_2.

Under this provision, the transition function f is defined as the function

$$f: I \times S \rightarrow S,$$

or, expanding I according to its definition,

$$f: (I_1 \times I_2 \times \ldots \times I_k) \times S \rightarrow S.$$

To discuss particular actions of the mechanism f at a particular time t, we require a notation that names the mechanism's state and its input states *at that time*. We can do this by extending our notation slightly. Let $S(t)$ designate the state of the mechanism at time t, and $I_j(t)$ designate the state (input value) of input j at time t. Then the mechanism's *dynamic*—its behavior over time—is specified by f according to the following formula:

$$S(t+1) = f(I_1(t), I_2(t), \ldots, I_k(t), S(t)).$$

That is, f uses the value of the state $S(t)$ of the mechanism at time t, and its inputs $\{I_1(t), I_2(t), \ldots, I_k(t)\}$ at that time, to determine the mechanism's state $S(t+1)$ at the next instant of time, $t+1$. We can apply f again, given the inputs $\{I_1(t+1), I_2(t+1), \ldots, I_k(t+1)\}$ at time $t+1$, to yield the state $S(t+2)$ at time $t+2$, and so on for times $t+3$, $t+4$, ... This iteration of f yields successive states, the so-called *state trajectory* of the mechanism, under the influence of the sequence of input combinations $I(t)$, $I(t+1)$, $I(t+2)$, ... Such iteration is the hallmark of generating procedures.

We have not yet said what outputs a mechanism produces in response to an input sequence. It simplifies the notation, without changing the range of rule-governed systems that can be modeled, if we simply take the state of the mechanism to be its output, as if we could observe everything or anything that takes place in the mechanism. Of course, in realistic cases we may only be able to observe a part of the state, as when we can see the hands of a watch but not the internal works. That complication is easily handled when we provide for the interaction of mechanisms as we'll see in the next section. With this simplification the transition function f completely specifies the mechanism's behavior. In what is called an abuse of notation, the symbol f will be used as the mechanism's name, in addition to designating its transition function.

Interactions and Linkage

Our basic concern in defining constrained generating procedures is to provide a setting in which we can study the complexities, and examples of emergence, that arise when rule-governed entities interact. I've adopted mechanisms, with their defining transition functions, as the formal counterparts of rules. Emergence, in this setting, is closely connected to the *generated* complexity that arises when multiple copies of like mechanisms interact. Our concern now is to provide for the interaction of copies of a set of elementary mechanisms.

A little more carefully, the definition of a constrained generating procedure starts with the selection of a set of mechanisms F, called *primitives*, from which everything else will be constructed. The mechanisms in F are the counterparts of the elementary mechanisms the Greeks used to construct all machines. When the *cgp* is used to model a game, neural net, cellular automaton, or other system exhibiting emergence, the primitives are the elements that are connected to form the model.

In this precise setting, one mechanism is *connected* to another when the state sequence of the first mechanism determines a sequence of values for one of the second mechanism's inputs. Once

F is selected, we connect copies of mechanisms in F to form a network of interacting mechanisms, as when we connect neurons to form a neural network. In short, we obtain a particular *cgp* when we select a set of primitives F and then interconnect them.

Let F be a small set of m elementary mechanisms defined by the transition functions $f_1, f_2, \ldots, f_m$. The mechanisms in F may have distinct sets of states, different numbers of inputs, and different alphabets for each input. To indicate these possibilities, we add an additional index to the letters designating these components: $I_h = I_{h1} \times I_{h2} \times \ldots \times I_{hk(h)}$ designates the possible input combinations to mechanism h, and $k(h)$ is the number of inputs to mechanism h. With this addition the transition function f_h for mechanism h has the form

$$f_h: I_h \times S_h \to S_h.$$

To complete the definition of *cgp*'s, it only remains to provide a way for (copies of) the mechanisms in F to interact. For two mechanisms to interact, the state of one mechanism must somehow determine one of the input values of the other mechanism. The mechanisms are then coupled or *connected,* much like a train of gears in a watch. Because the mechanisms have different state sets and different input alphabets, we need an *interface* to translate the state of one mechanism into a legitimate input (letter) for the other.

Once more, we define this idea precisely with the help of a function, the *interface function.* We can make it easier to define the interface function by collecting the different state sets of the mechanisms in F into a single set $S = S_1 \cup \ldots \cup S_m$. Because S is made up of the union of several sets, it is generally larger than any component, such as S_1. With this formal arrangement we can use S as the argument for all the interface functions. (For the purist, we map the forbidden states for some f_h into a dummy value, which in effect says "not allowed.")

Because of the different input alphabets and state sets, we must associate an interface function g_{ij} with each input j of each mechanism i. The g_{ij} produces a legitimate value (letter) for the input j of mechanism i, using the state of the mechanism connected via input j as its argument. Thus, g_{ij} has the form

$$g_{ij}: S \to I_{ij}.$$

We say mechanism h is *connected* to input j of mechanism i when, for all times t,

$$I_{ij}(t) = g_{ij}(S_h(t)).$$

That is, the input letter on input j at time t is determined, via the interface function, from the state $S_h(t)$ of mechanism h at time t. An input that is *not* connected is called *free*. For a free input, the values at each time must be supplied exogenously (from outside the *cgp*). In effect, free inputs amount to inputs to the whole *cgp*.

To give the *cgp* full scope for constructing models, we must capture all the different ways of connecting copies of the mechanisms in F. The easiest method is to specify how more complex *cgp*'s can be built up from simpler *cgp*'s. We'll start with the simplest *cgp*, a single mechanism, and then work our way up.

1. A *cgp* C can consist of a single mechanism $f \in F$.
2. Let C be a *cgp* already constructed, and let mechanism i in C have a free input j. Then the result of connecting input j to some other mechanism h in C (a new connection, from h to i, inside C) gives a new *cgp* C'.
3. Let C_1 and C_2 be *cgp*'s already constructed, and let mechanism i in C_1 have a free input j. Then the result of connecting input j to some mechanism h in C_2 (so that input j is no longer free) gives a new *cgp* C''.
4. All *cgp*'s based on F are formed via combinations of the previous three steps.

Use $n(C)$ to designate the total number of mechanisms in *cgp* C. (Recall that each of these mechanisms is a copy of some mechanism in F.) It will be useful to assign a unique index (address) x, drawn from the set $\{1, 2, \ldots, n(C)\}$, to each mechanism in the *cgp* C. We can do this by building up the indices in the same way we built the *cgp*.

1. When a *cgp* C consists of a single $f \in F$, the index $x=1$ is assigned f.
2. When a *cgp* C' is formed by connecting a free input in C to some mechanism in C, the indexing remains unchanged.
3. When a *cgp* C' is formed from C_1 and C_2 by identifying a free input in C_1 with a mechanism in C_2, the indexing of C_1 remains unchanged and each index x in C_2 is increased by $n(C_1)$ to yield a new index $x'=x+n(C_1)$. Then $n(C')=n(C_1)+n(C_2)$.

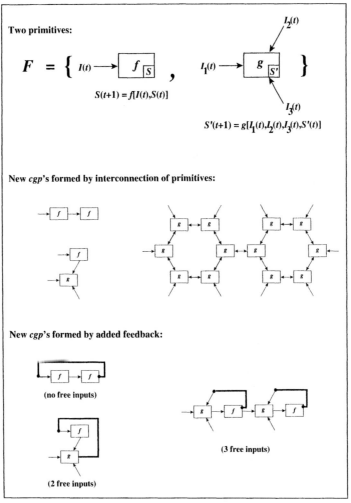

FIGURE 7.2

Constructing *cgp*'s from a set of primitive mechanisms.

Once each mechanism in the *cgp* is assigned an index, we can develop a kind of lattice to illustrate the interconnections and neighborhoods in the *cgp*, a lattice reminiscent of the tree that showed how moves are interconnected in a game. Node *i* in the lattice corresponds to the mechanism with index i. If mechanism i is connected to mechanism j, then an arrow runs from i to j in the lattice. For example, the mechanisms might be connected in a regular square array, like a checkerboard. Then each node would have four neighbors to which it is connected, and the whole array could be laid out as a square tiling pattern. However, any kind of connection pattern is allowable; it need not be regular in any way. We can even allow the number of nodes (copies of mechanisms) to become infinite, yielding an infinite array such as a checkerboard extending indefinitely in two directions.

Cellular Automata as cgp's

We can make an early test of the ability of *cgp*'s to model complex systems by looking at *cellular automata*. Cellular automata are the brainchildren of two of the most renowned mathematical physicists of the twentieth century, Stanislaw Ulam and John von Neumann. Ulam's idea (see the 1974 collection of his papers) was to construct a mathematically defined model of the physical universe within which one could build a wide range of "machines." This model physics would retain the essence of a real physics—it would have a geometry and a set of locally defined laws (a transition function) that held at every point in that geometry. Ulam saw that you could use a checkerboard geometry with an identical finite set of states assigned to each square (see Figure 7.3 for a simple example, Conway's automaton). Ulam completed the construct by specifying a transition function that determines the way the states change over time. Von Neumann then used this construct to design the self-reproducing machine mentioned in Chapter 1. Cellular automata have proved to be useful tools for investigating complex systems, so they constitute an appropriate test for *cgp*'s.

Each square, or *cell*, in the cellular automaton, with its associated states and rules, becomes a mechanism in the *cgp* framework. That is, the states of the cell become the mechanism's states, and the rules of the cellular automaton define the mechanism's transi-

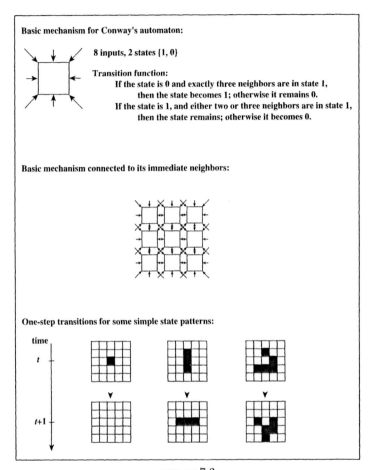

FIGURE 7.3
Conway's automaton.

tion function. In these terms, the cellular automaton is characterized as a *cgp* employing copies of a *single* mechanism, connected to form the regular square array described in the previous section. By picking an appropriate pattern of states in this array, we can provide a kind of active blueprint for any conceivable machine.

A Simple Cellular Automaton

One of the simplest cellular automata was designed and named by John Conway. It is called "Life." Life, despite its simplicity, offers some fascinating examples of emergence. The automaton will be fully defined here, and we will discuss one of the examples of

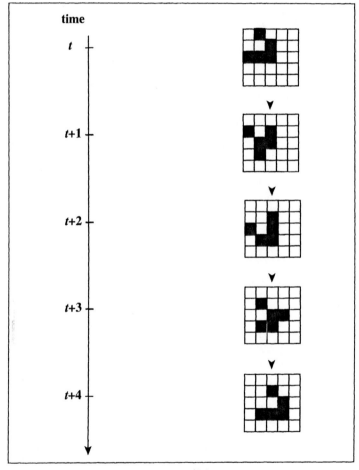

FIGURE 7.4
Successive transitions of the glider state pattern in
Conway's automaton.

emergence therein in some detail, but the reader should see Gardner's 1983 book to gain an idea of the extensive research that has been devoted to this automaton. In *cgp* terms, Life is formed from copies of a single mechanism that has just two states, represented by 1 and 0. For descriptive purposes, we can think of a cell as *occupied* by a particle if the corresponding mechanism is in state 1; otherwise it is *empty*. The lattice of connections is a square array, as described earlier, but now all eight squares that surround a given square are connected to that square. That is, each node has eight immediate neighbors (see Figure 7.3).

The transition function is also easy to describe. If a cell is empty (state 0), and exactly three of its eight immediate neighbors are occupied (state 1), then that cell becomes occupied on the next time-step (its state changes to state 1); otherwise, it stays empty. If a cell is occupied, and either two or three of its immediate eight neighbors are occupied, the rest being empty, then on the next time-step the cell remains occupied; otherwise, it becomes empty. That encompasses the full range of possibilities.

Such a simple *cgp* would seem to offer little that is instructive concerning emergence, but that is not the case. The first surprise is that you can embed a general-purpose (fully programmable) computer in this simple space. Specifically, you can design a pattern of occupied cells in this space that interact so as to form a general-purpose computer. In consequence, any process that can be modeled on a computer can be imitated in the "physics" of Conway's cellular automaton. I will not attempt a description of the complicated "general-purpose computer" pattern here, but will instead describe a much simpler example of emergence in Conway's automaton.

Gliders

Conway's automaton can contain a simple, mobile, self-perpetuating pattern called a *glider*. It consists of a pattern of five occupied cells, surrounded by empty cells (see Figure 7.4). The transition function just described produces a sequence of changes in the pattern over successive time-steps. Though five occupied cells are always involved, the pattern changes shape in a regular way and it moves ("glides") diagonally across the space. At intervals of four time-steps, the pattern recurs, now shifted one cell diagonally, say down and to the right, as in Figure 7.4. If you continue on in time, the pattern continues to go through the same sequence of changes, gliding downward across the lattice, as long as it does not encounter another pattern of occupied cells. The transition is so easy that you can try it on a piece of ruled paper.

Simple though the glider is, it tells us much about emergence. First of all, the glider is *not* a fixed set of particles bound together and moving on a trajectory through the space. Rather, particles are continually being created and deleted to produce the glider. This persistence of pattern, despite a continual turnover of constituents, recalls the persistence of the standing wave that forms in front of a rock in a rapidly flowing stream. The glider is more complicated than the standing wave because it changes form as it moves, and it does not stay in one location. Still, the glider and its glide are a consequence of the simple laws of this universe.

The possibility of such a spatially coherent moving pattern is not something easily determined by direct inspection of the laws of Conway's universe. The possibility only exists because of the strongly nonlinear interactions of the particles (states) in adjacent cells. Even if we restrict attention to the five-by-five array of cells sufficient to contain a glider, along with the immediate surround of empty neighbors, no extant analytical technique will predict the existence of a glider pattern. We can only discover the glider by observation, watching the laws play themselves out in different configurations. A five-by-five array in Life has $2^{25} \sim 32,000,000$ distinct configurations, so the task is not easy.

If we undertake an empirical exploration, we soon find out that many configurations are ephemeral, lasting only a few time-steps before dissolving to a set of empty cells. Other configurations remain fixed in place, either retaining a fixed form or "blinking" through a recurrent sequence of changes. Still other configurations migrate over long periods, assuming a multiplicity of forms, only to dissolve or wind up as a fixed configuration. The glider fits none of these categories, because it persists indefinitely while moving through empty space.

Even though we may infer the persistence of the glider from extended observation, such inferences can be treacherous. For instance, there are patterns in Life that go through extremely long sequences of changes before ultimately dissolving. We can only be sure of the glider's persistence if we can *prove* this property within a theory founded on the laws that define Life. In this, Conway's artificial universe is not so different from the universe in which we live. Both require a combination of experiment and theory to discover and explain regularities.

Once we establish the glider's persistence, we can think about it as a component of larger, more complicated patterns. In fact, this step was taken in showing that a general-purpose computer could be embedded in Life. In an embedded computer, the gliders serve as signaling devices, conveying information from one part of the pattern to another. Additional research showed that there are configurations that emit gliders, and others that respond to a collision with an incoming glider. These are the basic elements for the creation, transmission, and reception of information. Once information can be transferred from point to point, it takes only some simple "bit flipping" and storage schemes to provide computation. Thus, the existence of the glider encouraged a search for a way of constructing a general-purpose computer within Conway's universe. The ensuing search, and proofs, were relatively straightforward once conceived, but it is unlikely that they would have been pursued with the necessary intensity without this encouragement.

Lessons about Emergence

What lessons about emergence can we draw from this example? We see that:

1. Rules (transition functions) that are almost absurdly simple can generate coherent, emergent phenomena.
2. Emergence centers on interactions that are more than a sum-

ming of independent activities (imposed by the nonlinear rules in this case).

3. Persistent emergent phenomena can serve as components of more complex emergent phenomena.

All these points have been made earlier, but here they appear in a context so stark that nothing lies hidden in the complexity, nor is there any room for mysterious, unexplained activity.

CHAPTER 8

Samuel's Checkersplayer and Other Models as Cgp's

In CHAPTERS 3 AND 4 we closely examined two modeling projects that increased our understanding of learning and self-organization: Samuel's checkersplayer and models of Hebb's theory of the central nervous system. Both studies took place at IBM at the dawn of the computer age, but they teach lessons that are surprisingly current. We have only recently advanced beyond what was learned at that time, and many current studies fail to take those lessons about learning and self-organization into account. In part, such failures occur because the field moved in other directions for a long time. But the lessons also seemed largely tied to the specific efforts, checkers and Hebb's theory. Their general applicability was questioned.

Past chapters have highlighted the general principles extracted from these efforts. Nevertheless, the point remains that these projects were *very* different in implementation, despite similar aims. Can they be compared, and do the principles hold in a more general setting? Here we have an opportunity to try out the framework provided by constrained generating procedures. If we can translate both efforts into that framework, we have an opportunity to see what they hold in common.

I'll start with the checkersplayer, going into some detail about the steps required to embed it in the *cgp* framework. This procedure will pay substantial dividends when we come to examine the CNS model. We'll see that steps for embedding the CNS model are much the same, and that the differences are differences in detail. Then I'll look briefly at a recent, quite different, computer-based model that exhibits emergent phenomena, to see if the commonalities of the first two models hold up in this new context.

Samuel's Checkersplayer

Because the same rules apply to all squares of the checkerboard, it is natural to treat each of the thirty-two accessible squares as being controlled by the same mechanism. That is, we define a transition function (mechanism) f that specifies the rules of the game, then we form the *cgp* by interconnecting thirty-two copies of that mechanism. On a checkerboard a piece only moves to, or across, one of the four squares diagonally adjacent to the square it occupies, which suggests that each mechanism in the *cgp* should be connected to four neighboring mechanisms, in a regular diagonal lattice (see Figure 8.1).

In addition to implementing the rules, the *cgp* must allow the players to choose moves. That is, each mechanism must have a free input (an input *not* connected to another mechanism), with values that are set by the players. In the case of checkers, each mechanism thus will have at least five inputs. Four of the inputs will be connected to the mechanisms representing the diagonally adjacent squares. The other input will be the free input, which accepts information from outside the *cgp*.

To execute a move, we must use the free inputs to indicate where the move is to take place. In one simple approach, we use one free input to designate the square (mechanism) that is to be the origin of the move and another free input to designate the destination of the piece that is to be moved. The transition function is designed so that the states of the two mechanisms are changed accordingly. After the move, the state of the first mechanism indicates that it is now in the empty (no piece) state; the state of the destination mechanism indicates the presence of the piece moved.

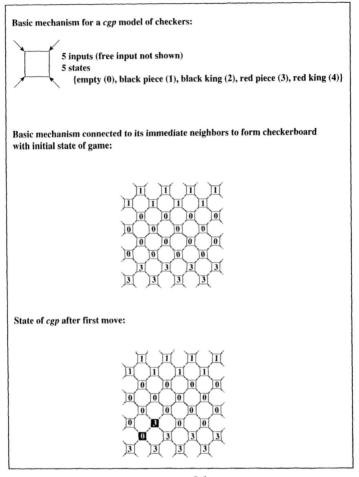

Basic mechanism for a *cgp* model of checkers:

5 inputs (free input not shown)
5 states
{empty (0), black piece (1), black king (2), red piece (3), red king (4)}

Basic mechanism connected to its immediate neighbors to form checkerboard with initial state of game:

State of *cgp* after first move:

FIGURE 8.1
A *cgp* for checkers.

Even this simple approach presents some interesting issues involving "action at a distance." For example, when there is a jump, the destination is not an immediate neighbor; squares intervening between the origin and the destination are affected. There are alternatives for solving this problem: we can enlarge the neighborhood of each mechanism, perhaps including all thirty-two mechanisms in the neighborhood, or we can send a "signal" through the intervening mechanisms to alert them to the impending move and its effects. A closer look at these alternatives gives us a better understanding of the options that *cgp*'s provide for constructing models of emergence.

There are several ways to use the free inputs in the *cgp* to indicate a player's choice of move. One of the simplest is to activate one free input at the point of origin of the intended move and another free input at the destination. More specifically, let mechanism (square) i be at the origin of the move and let mechanism (square) j be the destination of the move. In addition, let the index (address) of the free input be 1, so that I_{i1} and I_{j1} are the two inputs selected for activation. Under this arrangement we can limit the alphabets of the free inputs to just three symbols: {0 (inactive), 1 (origin), 2 (destination)}. Thus, if the move at time t involves mechanisms i and j, as origin and destination respectively, then $I_{i1}(t) = 1$, $I_{j1}(t) = 2$, and all other free inputs in the *cgp* are set to 0.

To specify the transition function f, we start by noting that each square is either empty or it is occupied by one of four pieces—a black checker, a black king, a red checker, or a red king. So we define S to contain five states corresponding to these possibilities, assigning the numbers {0 (empty), 1 (black checker), 2 (black king), 3 (red checker), 4 (red king)} to the respective possibilities. That is,

$$S = \{0, 1, 2, 3, 4\}.$$

We also know that only one piece can move at a time. An uncrowned checker has only two options, a simple move or a jump. (In checkers, if a jump is available, it is forced; the jump must be executed in preference to a simple move.) Let's look at each of these two options in turn.

1. In executing a simple move, the checker moves to an unoccupied, diagonally adjacent square in the forward direction. In the *cgp*, when

the move is executed, the mechanism corresponding to the occupied square goes to state 0 (unoccupied), and the state of the newly occupied square takes on the state of the originally occupied square (see Figure 8.1). To give an explicit example, let's assume that square i is occupied by a red checker, the move is to square j, and this is move t in the game tree. Then, before the move,

$$S_i(t) = 2, \text{ and } S_j(t) = 0.$$

After the move,

$$S_i(t+1) = 0 \text{ and } S_j(t+1) = 2.$$

All other states in the *cgp* remain the same.

2. The checker can jump over a piece on a diagonally adjacent square in the forward direction, if the square beyond, in the same direction, is unoccupied. In this case the move looks beyond the immediate neighborhood of the square initially occupied by the checker. To provide for this situation in the *cgp*, we have two broad options. (a) Enlarge the neighborhood so that each mechanism can be in direct contact with more distant mechanisms, or (b) make it possible for the mechanisms in the *cgp* to send and receive signals through intermediate mechanisms, so that they can test the state of mechanisms not in the immediate neighborhood.

At this point, we encounter the kind of tradeoff typical of most attempts at model building. Option 2(a) generates a more complicated connection pattern for the *cgp*—in the limit we might have to connect each mechanism to all of the other mechanisms in the *cgp*. We lose "locality" in the interactions. Option 2(b) requires a much larger state set and a more intricate transition function. Moreover, to allow for propagation of the signals, the *cgp* will have to allocate several time-steps to each move. Option 2(a), enlarging the neighborhood, is straightforward; but option 2(b), signaling, is worth a further look.

Signaling between Agents

Let's concentrate on signals as a way of providing interactions between mechanisms that are not immediate neighbors. Signaling is
one of the features that makes agent-based models complex, and it
contributes directly to the emergent properties of such models.
The very essence of an agent-based model is that individual agents
are in direct contact with only a limited number of other agents
(often only one other) at any given time. Thus, if large aggregates
are to be affected, the effects must be propagated from agent to
agent. The step-by-step nature of signal propagation, with attendant delays, adds greatly to the complexity of agent-based models.
And there is the oft-realized possibility that the signal is fed back to
the initiator, after going through several modifications en route.
We've already seen two illustrations of this effect in Chapter 5: the
feedback of signals in *exclusive or* circuits, and the reverberation of
neural nets with cycles. In the first case, we get the emergent phenomenon of indefinite memory, and in the second case we get the
emergent phenomena of cell assemblies and anticipation. Signaling must be readily accommodated in *cgp*'s if they are to serve as a
useful framework for studying emergence.

When we deal with models using signal propagation, there is
typically a transition from local to global (trees to forest) as we gain
experience. At first we concentrate on the myopic view provided by
local signaling laws and transitions. For example, while we're first
learning a game such as checkers, we concentrate on the movement possibilities of individual pieces. These are the trees. Then we
gradually see formations and other aggregate effects. Those are the
forest. It is only when we begin to see the emergent properties of
the forest that we begin to comprehend the whole.

The signaling technique we'll examine here is useful for other
kinds of signal-processing models, because it provides a way of obtaining "action at a distance" (see Misner et al., 1970). In particular, signal processing is critical in a *cgp*, where all interactions are
local. We'll retain the original neighborhood of the checkersplayer, the four diagonally adjacent squares, requiring the signal
to propagate square by square. With this arrangement it will take

the signal several time-steps to propagate to its destination, and several more time-steps for a return signal to indicate the condition of the destination. For example, if we are contemplating the legality of a jump, we have to check whether the second square in the direction of the jump is unoccupied. It will take two time-steps for the signal to propagate to that square, and another two time-steps for a signal to return indicating whether or not that destination square is occupied (see Figure 8.2). The technical details follow.

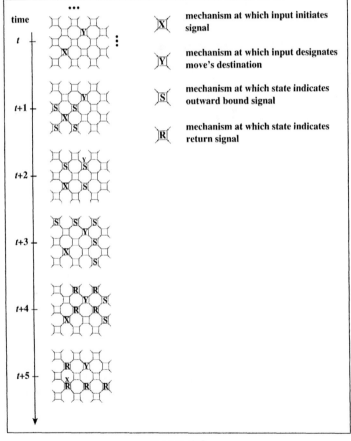

FIGURE 8.2
Signal propagation.

To make signal propagation possible, the mechanism must keep track of whether or not a signal is currently passing through a given point in the cgp. To do that we have to enlarge the state set of the mechanism. The transition function of the mechanism must also be able to read this information and act accordingly. To make this possible, we enlarge S so that it now has two components, R' and S', where $S'=\{0, 1, 2, 3, 4\}$, recording the piece information as before, and R' records the signal processing status. That is,

$$S = R' \times S',$$

so that the state of the mechanism is given by a pair of numbers. Under this provision the transition function has the form

$$f: (I_1 \times I_2 \times \ldots \times I_k) \times (R' \times S') \rightarrow (R' \times S').$$

If only one signal were passing through a mechanism at any given time, we could set $R'= \{0 \text{ (signal present)}, 1 \text{ (signal absent)}\}$. Then, for example, when there is a red checker on square i and no signal passing through, the state would be

$$S_i(t) = (R'_i(t), S'_i(t)) = (0, 3).$$

To move that red checker, we would set $I_{i1}(t) = 1$, and f would change the state of i so that, at time $t+1$,

$$S_i(t+1) = (R'_i(t+1), S'_i(t+1)) = (1, 3).$$

While the basic idea is simple enough despite the notation, there are complications. Further modifications are necessary to take these complications into account. The details have a certain interest, but they take up a fair amount of space, and they are not necessary to the topics that follow. I fall back on the well-worn phrase of textbooks: it makes a good test of comprehension to fill in the details as an exercise. Here I will only list the complications and suggest the resolutions, without going into detail.

1. There must be return signals to the origin showing that the necessary conditions for a jump have been met—a piece on an adjacent square, and an unoccupied square just beyond it.

For the signal to propagate outward, each of the neighbors connected to i "reads" the new state of i, through the input connected to i. The transition function at each immediate neighbor uses this input, the new value of i's state, to change the state of R' at the neighbor to 1. Thus, if j is one of the diagonally adjacent neighbors of i, $R'_j(t+2) = 1$.

If the signal is to propagate outward in an ever-widening circle, mimicking the ripple that spreads when a pebble is dropped into a pond, then each point just occupied by the signal must be reset to 0 after the signal passes through. In the notation just used, this means that

$$R'_i(t+2) = 0.$$

For example, if j, k, and l are neighbors at distances 1, 2, and 3 from i, then a ripple starting at time $t=24$ would show up as follows:

$$R'_i(24) = 1, R'_j(24) = 0, R'_k(24) = 0, R'_l(24) = 0$$
$$R'_i(25) = 0, R'_j(25) = 1, R'_k(25) = 0, R'_l(25) = 0$$
$$R'_i(26) = 0, R'_j(26) = 0, R'_k(26) = 1, R'_l(26) = 0$$
$$R'_i(27) = 0, R'_j(27) = 0, R'_k(27) = 0, R'_l(27) = 1.$$

Because the R'-component only stays equal to 1 for one time-step, as the signal passes through, the reset is easy. If an R'-component is 1 at some time t, then f automatically sets it to 0 on the next instant, $t+1$.

This is where the input signal at the destination comes in. When the ripple from the origin meets the mechanism designated by its input as the destination, a "return" signal is propagated. This signal indicates whether or not the destination is occupied. The state set must be further enlarged to accomplish this.

2. It is possible that some illegal move has been attempted. For example, there could be an attempt to move two pieces at once, by using the free inputs of two different mechanisms to designate the origin of two different moves.

The spreading ripples from the *two* points of origin provide a means of forestalling such illegal attempts. Roughly, when the two ripples intersect at some mechanism, as they must, we have at that mechanism an overt indication of the illegal attempt. Because no such intersection will occur if there is but one designated origin, such an intersection is diagnostic of an illegal attempt at two simultaneous moves. The mechanism at the intersection can then send out a signal blocking the attempt. This requires another enlargement of the state set.

3. There are other complications, too, such as multiple jumps, kings that can move both forward and backward, and "kinging" in the last row.

   The way to handle these complications is to enlarge the signal alphabet again, so that different kinds of signals can propagate, and be distinguished, simultaneously.

I have discussed this signal-propagating version at some length, because signal propagation allows "locality." Locality (keeping the interconnections in a *cgp* limited to a small neighborhood) plays an important role in model building of all kinds. Questions of locality become central where "action at a distance" has a major role, as in chess. (In chess a piece such as a bishop or a rook can move an indefinite distance if its path is clear, constrained only by the edge of the board.) In physics, models rise or fall on this point of locality; witness the ongoing debate about locality in the models proposed by quantum theory (Jammer, 1974).

Standing Back

This discussion has concentrated on designing a constrained generating procedure that captures the rules of a game. We've seen how a mechanism's states can summarize all that is going on at a given location in the game. With this provision, the mechanism's transition function gives a precise way of presenting the rules and constraints that define the game. The next step is to provide interconnections that mimic location adjacencies. The last step is to design the transition function so that the mechanisms can only interact in ways that yield states corresponding to legal game configurations, the states of these mechanisms being constrained according to the rules of the game. Furthermore, changes in these states, initiated via the free inputs, are constrained to correspond to legal moves. Thus the *cgp* generates the state (game) tree that is the hallmark of the game.

It remains to determine exactly what moves are made during a particular play of the game. That is the role of the checkersplaying program itself. To add Samuel's checkersplayer to this game-defining *cgp*, I would have to go on to design another part of the *cgp* that plays the checkersplayer's role. It would have mechanisms corresponding to feature detectors—mechanisms that would respond to states of the checkerboard *cgp*. It would also have arithmetic mechanisms (adders and the like) to provide an evaluation function. And this new *cgp* would have connections to the free inputs of the checkerboard *cgp*, to allow the new *cgp* to execute its move selections. All of this would be an interesting, rather intricate exercise, but it would not add substantially to our exercise in the use of *cgp* notation. Indeed, the exercise has already revealed some of the most important points about modeling checkersplaying as a *cgp*.

First of all, the *cgp* framework shows that there are hidden complexities, even in a game as simply defined as checkers. The *cgp* does *not* restrict us to one canonical model of the game. Instead there are significant alternatives. We can choose an extended neighborhood and nonlocality, or we can commit ourselves to signal propagation and small neighborhoods. Each of these alternatives has advantages and disadvantages. Questions that are easy to approach in one model can be quite difficult in the other. As we go along, we'll learn more about such choices.

We've also seen that building into the *cgp* a distinction as to what is legal and what is not is not as obvious as it might seem. It's clear enough from the rules of checkers that only one piece can be moved at a time, but to embed this restriction into the *cgp* takes some thought. We could, of course, put the proscription on the "user," saying that the user must not select more than one free input at a time to designate the origin of a move. But that is really not satisfactory, because it puts the proscription outside the definition of the *cgp*. Rather, the *cgp* should be designed to make it *impos-*

sible to execute simultaneous moves. In the signal-propagating version for checkers, this took some thought about using the intersection of the two ripples, spreading out from the two points of origin, to prevent the illegal action. This is a kind of exclusion principle, slightly reminiscent of Pauli's exclusion principle in physics (Feynman, 1964), that requires the component mechanisms in the *cgp* to detect and avoid illegal configurations.

Finally, we've seen again that the interaction of simple mechanisms can generate a very complex range of possibilities. That, more than anything else, has been the point of this exercise. Within a *cgp* we can recapture the branching, complicated game tree that gives rise to the perpetual novelty of board games. We'll see that this way of capturing complexity serves us well when it comes to investigating other models and modeling processes.

Central Nervous System Models

We can gain additional intuition about the relation between modeling and constrained generating procedures by looking at some rather different models. I will not pursue these as closely as the checkers model; instead, I will concentrate on a few salient points that emphasize other *cgp* facets. We spent considerable time earlier contrasting emergence in CNS models with emergence in Samuel's checkersplayer, so I'll start there. We can now reap some dividends from those comparisons, by looking at CNS models in the *cgp* framework.

The customary models of the CNS use a single mechanism, the modeled neuron. Copies of this mechanism are interconnected to form the neural network. The transition function that defines this basic mechanism, as we noticed in Chapter 5, is built up from a function that has formal similarities to Samuel's valuation function—a simple weighted sum of arguments. In the case of the model neuron this function has additional complications: a variable threshold, a fatigue factor that depends on the neuron's firing rate, and Hebb's rule for changing the weights. With these complications the state of a neuron is determined by:

1. whether or not the neuron has fired at time t (which determines the threshold level);
2. what the current values of its weights w_i are (as determined by Hebb's rule);
3. what its current firing rate is (from which the fatigue can be determined).

The model neuron's transition function acts on information supplied by the neuron's neighbors (the presence or absence of pulses at the synapses of the neurons that connect to the given neuron). Using this information, along with the current state of the neuron, the transition function determines that neuron's next state, including new values for its weights and a revised firing rate. The overall action is much like signal propagation in the *cgp* version of the checkersplayer.

The whole comparison becomes more vivid if we attend to the formal presentation. The weighted sum of arguments is given by

$$\Sigma_i \, w_i \, v_i(s),$$

where the $v_i(s)$ are inputs from neighboring neurons, and the weights w_i determine the relative importance of each of those inputs in determining the neuron's behavior. A neuron fires (produces a pulse) when

$$\Sigma_i \, w_i \, v_i(s) - F > T,$$

where T is the variable threshold and F is the fatigue factor. We use this information to design a transition function in the usual form,

$$f: (I_1 \times I_2 \times \ldots \times I_k) \times S \to S,$$

for a neuron with k inputs, where each I_i corresponds to the $v_i(s)$ for the ith neighbor, indicating whether or not the state of that neighbor specifies an outgoing pulse. This information is used by f, along with the current state $S(t)$ of the neuron, to determine the neuron's next state, including new values for its weights and a revised firing rate.

In these *cgp* models each copy of a mechanism acts as a simple agent interacting with its designated neighbors. In the checkers-player, the agents are the squares of the board, with the state reflecting the kind of piece present on that square; in the CNS model, the agents are the neurons, with the state indicating the condition of the neuron. Unlike the checkerboard and cellular automata, with their regular arrays of interconnections, the CNS has an irregular array of interconnections, with much looping and feedback. This configuration does not make much of a change in the formal appearance of the *cgp*. It simply means that the lattice of interconnections between agents has a more intricate definition.

The *cgp* version of the CNS model teaches us some of the same lessons about emergence that we learned from the *cgp* version of the checkersplayer, despite the substantial differences in the two: the apparatus of mechanism (state and transition function), generators (rules), and agents again serves us well in capturing the model's performance. The locality, enforced by a limited number of neighbors for each agent, emphasizes signaling. The kinds of feedback that arise because of signaling enforce important constraints on the range of possibilities. These constraints give rise to complicated game trees (state transition diagrams) that nevertheless exhibit exploitable regularities.

The *cgp* version of the CNS model also provides one of those bonuses we expect of interdisciplinary comparisons, by clarifying some important properties that are obscure in other contexts. The emergent properties of synchrony, persistence, and indefinite memory, typical of CNS models with loops, give rise to the cell assemblies that determine an evolving "strategy" for exploiting regularities in the system's environment. The enhanced persistence and hierarchies that result have a critical role in the behavior of the original CNS models, and so it must be for the *cgp* versions of these CNS models. That, in turn, makes it likely that enhanced persistence and hierarchy will have a significant role in other *cgp* models, and in the study of emergence. We will see in Chapter 10 that this is so.

Copycat

I would be remiss if I did not mention another major computer-based modeling effort that directly confronts emergence: Hofstadter and Mitchell's "Copycat" (see Mitchell, 1993) model of analogy, one of several models produced by the Fluid Analogies Research Group under Douglas Hofstadter's direction. Copycat is an intricate, subtle model, to which a full chapter could be devoted. Indeed, the presentation here would gain in depth from such an exposition, but at the cost of a whole new range of detail. Rather than do that, I simply highlight a few of Copycat's features that bear on emergence in the *cgp* setting, urging the interested reader to pursue the topic by reading *Fluid Concepts and Creative Analogies* (Hofstadter, 1995).

Copycat is designed to reflect the fluid rearrangements that go into the construction of analogies. It is designed to work on letter strings of the type so familiar from intelligence tests: "Suppose the letter-string **abc** were changed to **abd**; how would you change the letter-string **ijk** in the same way?" Though simple, such problems involve a subtle mix of classification and recombination. Copycat's repertoire of possibilities is determined by a fixed network called a *slipnet.* The nodes of the slipnet represent the basic concepts (classifications) available to Copycat, while the connections represent the relations between these concepts.

We can think of analogy making in Copycat as proceeding from a source to a target. The source gives an instance that is to be used to guide an analogous completion of the target. In the example just given, the source is the pair of strings **abc** and **abd,** and the target is the string **ijk** with its to-be-supplied complement. This source/target interpretation of analogy will prove useful later when we discuss metaphor and innovation.

Though the slipnet is fixed in form, each of its connections has an attached number, called its *length,* and this number varies as the model is executed. As might be expected from its name, the length reflects the closeness of association between related concepts *for the problem at hand.* As Copycat explores the problem, the lengths change to reflect Copycat's varying estimates of the relevance and interrelatedness of its basic concepts vis-à-vis parts of the problem.

The exploration is carried out by a variety of simple agents called *codelets*. Agents create various kinds of groupings within, and bridges between, the source and target. The agents are directed by selected nodes in the slipnet; each time an agent successfully executes an action, the directing node receives a "jolt of activation." This activation spreads to nodes that are currently nearby in the slipnet, in amounts determined by the lengths currently assigned to the connections between nodes. Newly activated nodes bring new agents into play, and the spreading activation also causes changes in the lengths assigned to some connections.

The interactions in Copycat are more diverse and more intertwined than this brief summary indicates. For example, the connections between concepts are themselves labeled with concepts. These connection-labeling concepts participate in all the activities so far described, including activation; the level of activation of a label is the factor that determines the length of the connection. Each additional mechanism and interaction is well motivated by deeper considerations—there is very little in Copycat that is ad hoc. Because of this depth, I suspect that Copycat will become a classic, quite comparable to Samuel's classic work.

The net effect of the interaction of the slipnet and the codelets is to discover similarities between parts of the source and parts of the target (such as similar groupings), while simultaneously trying out ways of fitting these parts into comparable configurations. Partial matches, through spreading activation, bring new codelets into play, causing exploration of nearby concepts (nodes) in the slipnet, in an attempt to extend the comparisons. When the comparisons reach a sufficiently high level, the unfinished part of the target is completed via a bridge to a comparable group in the source.

Copycat exhibits a feature that will play an essential role later, in our discussion of metaphor and innovation. Spreading activation produces a kind of aura of association (Hofstadter calls it a halo) around any node. This aura changes with changing levels of activation and changing connection lengths, so there is no easy way to represent it as a list, or in some other deterministic fashion. The changing aura is a major contributor to the fluid nature of emergent analogies in Copycat. We'll see similar auras, of meaning and technique, play a key role in scientific model building, metaphor construction, and innovation.

Much that goes on in the interaction between the slipnet and the codelets involves computations similar to the computations used in modeling nerve nets with internal feedback loops. The level of activation of a node in the slipnet is the counterpart of the firing rate of a neuron in a neural net; the length of a connection is the counterpart of a synaptic weight; the decay of activation is the counterpart of fatigue; and so on. The agents (codelets) add another factor, but they can be mimicked by appropriate feedback loops in the neural net.

One could translate Copycat into the *cgp* framework along these lines, paralleling the *cgp* version of CNS models, except for one additional factor. Copycat invokes its agents in a stochastic (probabilistic) fashion: the level of activation of a node is used to determine the *probability* that an agent will be invoked. There is always the chance that some agent with a low level of activation will be invoked by a node. Thus, "long shots" can be tried on occasion, which would never occur if agents were invoked only by the nodes with highest activation. As a result, Copycat has an emergent architecture wherein "top-level behavior emerges as a statistical consequence of myriad small computation actions," to use Hofstadter's (1995) phrase.

To this point, there is no direct provision for probabilistic selection in a *cgp*. One *could* introduce a pseudo-random-number program into the *cgp*, using our ability to structure parts of the *cgp* as a general-purpose computer. We could then use these random numbers to implement the stochastic actions in Copycat (as was done in the original computer-based implementation). However, this would be a roundabout approach in which the *cgp* framework would do little to enhance our understanding. It is also true that the active agents in Copycat, the codelets, are only indirectly captured by the in situ mechanisms of a *cgp*. What this suggests is an enlargement of the *cgp* framework to allow a more direct account of mobile, stochastic agents. That is the subject of the next chapter.

CHAPTER 9

Variation

WHEN I EXAMINED the commonalities of examples of emergence in Chapter 6, I devoted a section to agent-based models. The aggregate properties of interacting agents, such as the ants in an ant colony, provide some of our best examples of emergent phenomena. Agents, above all, are mobile. And that mobility raises a question about the facility with which constrained generating procedures can handle agents. We know, from the example of the glider in Conway's automaton, that *cgp*'s can accommodate mobile objects. However, that is not the same as designing a *cgp* to handle agents according to specifications given at the outset.

A *cgp* with fixed connections works well when the focus is on the fixed geometric constraints of an underlying "physics." For mobile agents, the focus shifts to the making and breaking of connections. Then a fixed-connection *cgp* is both less convenient and less instructive. This chapter concentrates on changes in the formulation of *cgp*'s that allows them to directly accommodate changes in the geometry, placing those changes under the control of the *cgp* itself. As a result, mobile agents within the *cgp* can actually alter connections in the *cgp* to reflect their changing patterns of interaction. The examples at the end of the chapter will show that this extension is helpful in understanding the emergence that is a feature of agent-based models.

This chapter considerably extends our understanding of generating procedures, but it is the most technical chapter in the book. Its lessons are repeated later on, so the reader can jump over it without an impaired ability to understand later chapters; the cost will be depth, not breadth.

Cgp's with Variable Geometry

Allowing a *cgp* to make and break its own connections may sound paradoxical, requiring the system to have a plan that is then changed by the plan itself. However, as in the case of self-reproduction, there is a way to slip around the paradox. The solution in this case is related to the earlier solution. In the case of self-reproduction, the solution was to *generate* the overall plan from an initial description, in much the same way that the chromosomes in a fertilized egg are sufficient to generate the mature organism. Here the solution is to generate the connections indirectly, rather than listing them as we did for the fixed-structure *cgp*. In implementing this solution, we'll see that we can go a step further, making it possible for the *cgp* to add and delete mechanisms as well as change connections.

I'll use the acronym *cgp-v* (the *v* for variable structure) to designate this extension of *cgp*'s. The study of *cgp-v*'s is barely under way, but it is already apparent that it is a topic rich in structure and interpretation. Here I will only go far enough to illustrate the additions to our understanding of emergent phenomena offered by such *cgp*'s.

A *cgp* can change its own connections only if it has mechanisms, called *processors*, that can modify other mechanisms. A processor's transition function must accept other mechanisms as inputs, and its output must be in turn a mechanism. We can simplify the formalism if we design the transition function to process *descriptions* of mechanisms, rather than the mechanisms themselves. It is as if the transition function is fed designs as inputs, producing a new design in response (implementation is assumed to be automatic once the design is available).

To allow the *cgp-v* to use processors, we must enlarge the input alphabets for mechanisms to the point that they include descriptions of all the mechanisms in *F*. By introducing standard descriptions for those mechanisms, we can standardize the input alphabets. The use of standardized alphabets in turn standardizes the processing. That is the outline of the sketch that follows.

Tags

Tags provide an intuitive, convenient way of accomplishing modifiable connections. I examined tags at some length in *Hidden Order* (Holland, 1995), so I will only discuss the points directly relevant to defining a *cgp*-v.

Consider a *cgp* consisting of n mechanisms drawn from F, as defined in Chapter 7. But assume the mechanisms have not yet been interconnected. To define the connections in a *cgp*-v we first assign each mechanism an identifier or tag, called an *id tag* (see Figure 9.1). We can think of the id tag as a kind of address, which allows other mechanisms to access the given mechanism by knowing its associated tag.

To use this address, each input of each mechanism now has an associated *condition* that designates the tags of the mechanisms it will accept as input. The condition is a kind of filter. It looks over the n mechanisms in the *cgp*-v and decides which if any are suitable for that input; a mechanism is suitable if it has an id tag that is ac-

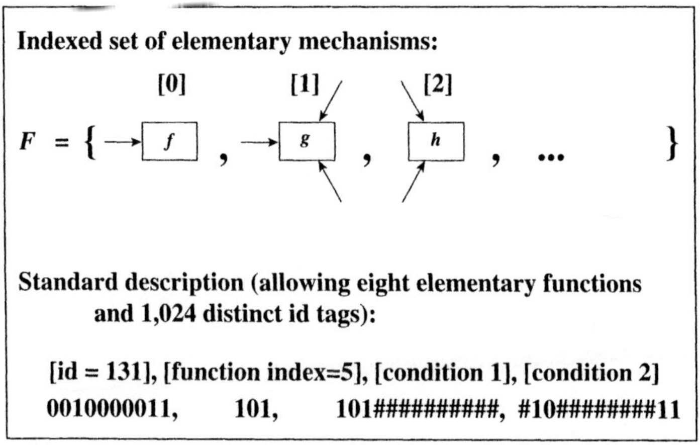

> **Indexed set of elementary mechanisms:**
>
> $$F = \left\{ \begin{array}{c} [0] \\ \rightarrow \boxed{f} \end{array}, \begin{array}{c} [1] \\ \rightarrow \boxed{g} \end{array}, \begin{array}{c} [2] \\ \boxed{h} \end{array}, \ldots \right\}$$
>
> **Standard description (allowing eight elementary functions and 1,024 distinct id tags):**
>
> [id = 131], [function index=5], [condition 1], [condition 2]
> 0010000011,   101,   101##########, #10########11

FIGURE 9.1
Standard description of a mechanism.

ceptable to the condition. Specifically, the condition is used as follows.

1. If there is no mechanism having an id tag that meets the condition, then the input is free—its values must be specified from outside the *cgp*-v, as in the case of a free input in a fixed *cgp*.
2. If there is exactly one mechanism with an id tag that meets the condition, then the input is "connected" to that mechanism— the letter in the input alphabet corresponding to the mechanism's description becomes the current value of that input.
3. If there is more than one mechanism with an id tag that meets the condition, then one of the mechanisms is selected at random (each time) to be connected to the input, the input value being determined as in 2.

These additions make it easy to change connections, because a connection between mechanism h and input j of mechanism i can be modified by changing either the id tag of h or the condition associated with input j of mechanism i (see Figure 9.3, page 169). However, to implement this procedure, we must have a set of descriptions for the mechanisms in F. Moreover, because the mechanism's state determines its output (the result of its processing), these descriptions must include the state of the mechanism. That is our next concern.

Standard Descriptions

Because a *cgp*-v is based on (copies of) a small set F of elementary transition functions, as was the case for fixed-connection *cgp*'s, we need only be concerned with the description of mechanisms in F. However, I will allow some simple variations on those mechanisms, mainly by allowing modifications of the id tags and input conditions. I'll also allow the *cgp*-v to change the number of (copies of) mechanisms it uses.

Because the makeup of the *cgp*-v can change, we must be able to list the mechanisms present at any given time. Let's call that list the *current-components list* (see Figure 9.2, page 166). According to our convention, the list will consist of descriptions of mechanisms. Also, according to that convention, each description that appears on the current-components list is automatically implemented. That is, each mechanism described on the list automatically becomes an active part of the *cgp*-v. To modify or add a mechanism, we simply modify or add a description to the current-components list (see Figure 9.3).

With these preliminaries in place, we can develop a standard way of describing the possible mechanisms. The augmented mechanisms we're dealing with have the following standard parts (see Figure 9.1):

1. an id tag;
2. a set of inputs, each with an input alphabet and an associated condition;
3. a transition function that uses the inputs and a set of internal states as arguments.

To provide a convenient, modifiable description of these augmented mechanisms, we must describe each of these parts unambiguously.

I'll use strings over a simple alphabet (three symbols will do) to encode or index the parts just described. This approach is reminiscent of representing numbers, indices, and instructions in a digital computer via binary strings, or describing different chromosomes via an alphabet of nucleotides.

The first step is to index the transition functions in the small fixed set *F*. Each of the indices can be encoded as a binary number. The id tags can also be limited to a finite number of distinct tags (this is not necessary, but it simplifies the discussion). A set of thirty bits allows a billion distinct tags.

Next, we encode the conditions associated with the inputs. I'll restrict the conditions so that they "look" only at the id tags and the function index. That is, the conditions ignore all other aspects of the mechanisms, basing their "decision" on the properties of the id tag and the function index. We can use a technique closely related to the standard description of rules in classifier systems

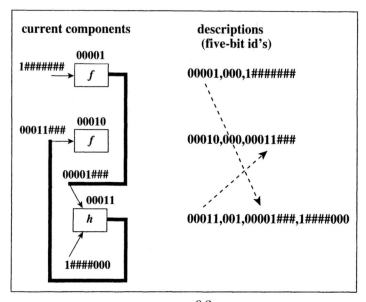

FIGURE 9.2
Mechanisms interconnected via tags and input conditions.

(see Chapter 2 of *Hidden Order*): the condition looks at the description string (the id and index part) bit by bit. At each position it either requires a particular bit value (a 1 or a 0), or else it "doesn't care" (it allows either value at that position). "Doesn't care" is indicated by a special symbol, #, in the condition string. Thus the condition string 1##0## requires a 1 at the first position of the description and a 0 at the fourth position, and allows any value at the other positions (see Figure 9.2).

With these provisions a mechanism is described by a string over the alphabet {1,0,#}. It is convenient to order the parts of the description so that the id tag is first, the function index is second, and the conditions associated with the mechanism's inputs (the transition function's arguments) are last. Under this arrangement the string has the form (id)(function index)(condition 1)(condition 2) ... (condition k).

To complete a formal definition of *cgp*-v descriptions, it only remains to produce a notation naming the parts. Let *G* be the set of indices for the transition functions in *F*, let *H* represent the set of allowed id tags, and let *C* represent the set of conditions. If the mechanisms in *F* all have two inputs, then the set of all admissible descriptions, *D*, is given by

$$D = H \times G \times C \times C,$$

using again the × notation we utilized earlier (in Chapter 7). An individual description $d \in D$, in this case, is the 4-tuple $(h, g, c1, c2)$, with $h \in H$, $g \in G$, $c1 \in C$, $c2 \in C$.

There are some further technical details, but they have no role in the discussion that follows. Suffice it to say that setting up easily parsed descriptions of the allowable mechanisms is a straightforward exercise. Moreover, the descriptions are reasonably compact. For instance, if the indices are 3 bits long (allowing 8 transition functions), and the id tags are 10 bits long (1,024 distinct tags), then the conditions will be described by strings of 13 letters, and the overall description (when there are two conditions) will be 39 letters long (see Figure 9.1).

Generalizing the Current-Components List

In the fixed-geometry *cgp*, we treated a mechanism's state as its output. Now I want to enlarge on that idea. The objective is to allow the output (state) to be treated as a description, thereby opening the option of using a mechanism to create descriptions of new mechanisms. To accomplish this we simply provide each state with a "header" (prefix) that is a description from the set of allowable descriptions. Accordingly, the state is a string with the form

header   state

(id)(function index)(condition 1)(condition 2) .. (condition *k*) (mechanism's internal state).

With this arrangement, the transition function for the mechanism can produce an output that has a changed description, because it is a part of both the function's domain (argument) and its range (value). In effect, the transition function can process descriptions.

The formal version of the addition of this header requires each state $s \in S$ to have the form

$$s = (d,s') \in S = D \times S',$$

where S' plays the role of the state set used with a fixed-connection cgp. Under this arrangement, the transition function for a cgp-v has the form

$$f: (I_1 \times I_2 \times \ldots \times I_k) \times (D \times S') \to (D \times S').$$

Because D is included in both the domain and the range, f can cause the description D to change.

We can exploit this potential for changing descriptions to the point that the cgp-v can modify its own structure. The first step is to place *all* the states (outputs) of all the cgp's mechanisms on a list much like the current-components list. Call it the *current-state list* (*csl*). All mechanism inputs, other than those provided by free inputs, must come from this list. That is, all interactions among mechanisms are mediated by this list.

Now we take a critical step: *we treat the current-state list as the current-components list.* Note that each entry on the list has a prefix drawn from the set of descriptions, so it is interpretable as a mechanism. Indeed, the entries on the *csl* provide complete information about the *cgp*-v, giving both a list of the mechanisms composing it as well as their current states. On each time-step, each entry on the *csl* is interpreted as a mechanism, with its action being dictated by the transition function indexed in the prefix. The result of those actions is the *csl* for the next time-step. This *csl* is executed in turn. And so it goes, time-step by time-step. The behavior of the *cgp*-v, with its changing connections and mechanisms, is fully determined by this procedure (see Figure 9.3).

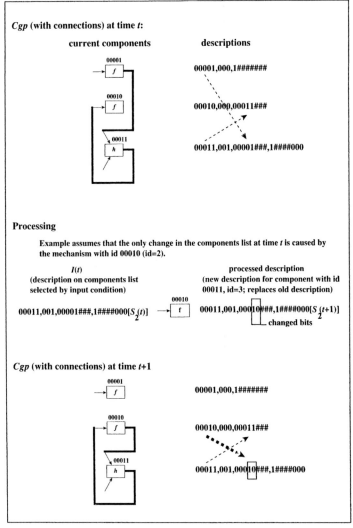

FIGURE 9.3
Internally controlled change of structure.

This self-referential procedure, where a string on the *csl* determines what will happen to itself, may seem confusing but it is fairly straightforward. The function index part of the string selects a transition function, and then that function is applied to the *whole* of the string (including the index part). Such self-referential action mimics what we do in a standard general-purpose computer. Every storage register in the computer contains a string of bits that is interpreted as a binary number. However, when the computer designates that storage register as the next instruction to be executed, that same number is also interpreted as an instruction. This duality allows the computer to manipulate its own instructions. One instruction can modify another by simply treating it temporarily as a number to be modified, using the usual arithmetic operations. This duality is exactly what gives a general-purpose computer its power.

A *cgp*-v gains a similar computational power under the convention that each entry on the *csl* is interpreted as a mechanism (instruction). In addition, the *cgp*-v satisfies the earlier convention that all mechanisms act simultaneously on each time-step, as in the case of cellular automata. This simultaneous action turns the *cgp*-v into a massively parallel computer. Every string on the current-state list is concurrently executed as a mechanism, whether or not it was put on the list for that purpose. Many of the "unintended mechanisms" will have states (outputs) that have no impact on the rest of the *cgp*-v; such descriptions serve as "numbers," with their transition functions and states being irrelevant to the rest of the *cgp*-v. There can even be dummy transition functions, with distinctive indexes, that "do nothing." The corresponding descriptions can then serve *only* as numbers, mimicking the part of a computer's store that is reserved for data.

Mechanisms as Processors

Note, again, that the *csl* now contains the *total* description of the *cgp*-v at any time, both its structure and its current state. When an entry on the list is interpreted as a mechanism, it uses its input conditions to select other strings on the list as its inputs. It can modify those strings, via its transition function, thereby producing new strings for the *csl*, strings interpretable as new mechanisms. In short, the mechanisms on the list can *process* other mechanisms on the list.

We gain a simplification if we standardize the local states S' for the different $f \in F$ that generate the *cgp*-v. This standardization mimics the standard-length strings in a general-purpose computer. In the general-purpose computer, any bit string can serve as the argument for any of the computer's instructions; in the *cgp*-v, this standardization means that all the *cgp*-v's arguments, input strings as well as local state, come from the same set. Because of the standardization, we no longer need the interface function that the fixed-connection *cgp* used (to translate mechanism states to mechanism inputs): each transition function $f \in F$ uses exactly the same set of states, $S = D \times S'$ for its inputs (arguments) and outputs (values), because S' is the same for all f.

Summing Up

The basic idea, then, is that the current-state list presents the complete description of a *cgp*-v at any time. It describes the mechanisms that currently make up the *cgp*-v, concurrently recording the states of those mechanisms. All entries on the *csl* are executed simultaneously.

In detail. Each entry on the current-state list has a standard format: (id tag) (transition function index) (input condition 1) (input condition 2) . . . (input condition k) (local state). As in the case of the fixed-structure *cgp*, each transition function f is drawn from a fixed, finite set F of allowable functions; because F is finite, the index set can be encoded by a string of fixed length. A similar statement can made for all other segments of the standard format, so descriptions in that format can be encoded as strings of fixed length over a small alphabet.

For each entry on the csl. The id tag serves as the entry's address;

the transition-function index specifies the transition function $f \in F$ associated with that entry (used when the entry is interpreted as a mechanism); the input conditions each accept a subset of the id tags; the local state serves any role the designer of the *cgp* desires. The value of each input, at each time, is determined by searching the *csl* for an entry that has an id tag satisfying the input's condition (see Figure 9.1).

To determine the action of the *cgp-v*, the following steps are executed (see Figure 9.3).

1. The condition part of each input of each mechanism on the current-state list is checked against the other entries on the list.

   a. If only one entry on the list has a tag satisfying the condition, then that string is assigned as the value of the input at that time.

   b. If several entries on the list have a tag satisfying the condition, then one of the strings so tagged is chosen at random as the value of the input at that time.

   c. If no entry on the list has a tag satisfying the condition, then the input value must be supplied from outside the *cgp-v*—the input is a free input. (Alternatively, a *null value* could be introduced, so that the transition function effectively ignores that input at that time.)

2. Once the input values for an entry have been determined, the entry's transition function $f \in F$ is applied to those values to determine a next state. The transition functions in F are so designed that their values are encoded as standard strings for the current-state list.

3. All of the (next) states computed by the different f are entered on the current-state list.

   a. If there is no string on the list with the same id tag, then the new string is added to the list; the string that produced the new string remains on the list.

b. If there is a single string already on the list with the same id tag as the new string, then that string is replaced by the new string. (It is sometimes useful to allow the output string to erase the *csl* string that matches its id, without replacing it. Erasure can be made conditional on the output string's local state.)

c. If there are several strings already on the list with the same id tag as the new string, then one of those strings is chosen at random for replacement (or erasure).

If a computed string (new entry for the *csl*) has a changed input condition or id tag, then the *cgp*-v's geometry of interactions changes. For example, if the input condition of a mechanism is changed, then it will typically attend to some new id tag, with the result that it will receive inputs from a different mechanism. Alternatively, if the id tag of a description is changed, then that description can satisfy different input conditions; it subsequently acts as input to other mechanisms.

Examples

Making a Cgp *into a General-Purpose Computer*

With an appropriate selection of transition functions for *F,* we can easily design a *cgp*-v that mimics a general-purpose computer: first, we designate a set of mechanisms to act as storage registers; they are identical except that each has a unique id tag. The state of each of these mechanisms plays the role of a stored number. To determine which of its stored numbers are to be interpreted as instructions, a general-purpose computer uses a special register, call it NEXT. That register indicates the address of a storage register containing the number that is to be decoded as the next instruction. That number is copied into another special register, call it INSTR, which decodes the number and executes the corresponding instruction. Finally, there is a register with special circuitry, call it ARITH, that carries out the arithmetic operations specified by the numbers successively decoded in INSTR. We can set up a distinct transition function for each of these special registers, supplying the corresponding mechanisms with id tags that identify them as INSTR, NEXT, and ARITH.

To give some idea of how this arrangement would work, let us look at the execution of a "store" instruction. In a typical general-purpose computer, when a number in a storage register is interpreted as an instruction, the interpretation has two parts. One part encodes the instruction type, "store" in the present case, and the other part encodes the "address" of the register to be used as an argument by the instruction. The typical "store" instruction copies the number in ARITH to the register specified by the address part, replacing whatever number was already there.

In the *cgp*-v, the state of the INSTR mechanism indicates the current instruction type and its address (see Figure 9.4). One of the inputs of the mechanism ARITH has a condition that accepts only the state of INSTR as input. Set the transition function of ARITH so that it uses the address part of INSTR's state to set the id tag of ARITH's next state. In particular, if INSTR indicates a "store" instruction, the "address" part of INSTR's state is assigned as the id tag of ARITH's state. Under the *cgp*-v rules, the string that is ARITH's state will be placed on the current-state list. Moreover, it will displace some string that has the same id tag. We design this *cgp*-v so that each mechanism corresponding to a storage register has a unique id tag, so ARITH's state (output) displaces a unique storage register string (number). The effect is to store the "number" in ARITH at the storage register mechanism addressed by the "store" instruction, as required.

One can mimic each of the operations in a typical general-purpose computer along similar lines, including updating the content of NEXT and INSTR on each time-step, to get the next instruction in the program from the storage registers. It is not a difficult exercise, but it takes some time. One can go even further, making *cgp*-v's that use multiple ARITH mechanisms to execute many instructions simultaneously. This leads to *cgp*-v's that are much more "parallel" but still have general-purpose capability.

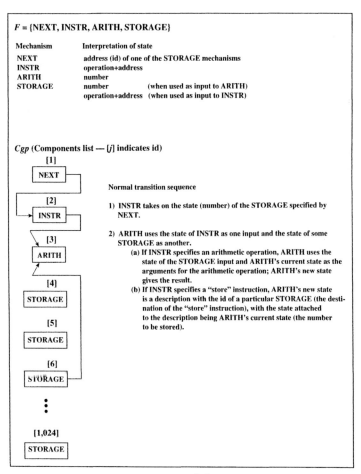

FIGURE 9.4

Configuring a *cgp* as a computer.

Making a Cgp *into a Cellular Automaton*

The use of id tags to provide a regular array of interconnections, as with a cellular automaton, nicely illustrates the addressing capability of tags.

We start with a single transition function, for instance, the rule for Conway's cellular automaton. Each mechanism in the array employs this function, but each has a different id tag to indicate its position in the array. Let's assume that the array of interconnections is square. Then we can specify the location of any point in the array by an (x,y)-coordinate scheme, giving its distance from a point of origin along horizontal (x) and vertical (y) axes. To provide these interconnections in the *cgp*-v, we simply encode the (x,y)-coordinate for each mechanism as a string of bits in its id tag. For example, if we're considering a 16-by-16 array we can encode each coordinate with four bits, requiring a total of eight bits for this purpose (see Figure 9.5).

Once we've made this id tag assignment, we set the input conditions so that each input accepts the id tag of one of the mechanism's immediate neighbors. For example, the mechanism with id tag (x,y) would have one input that accepts the id tag of the neighbor at $(x+1, y)$, another input that accepts $(x+1, y-1)$, and so on around the square of eight neighbors. With this arrangement, the

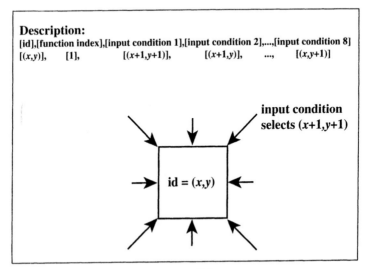

Description:
[id],[function index],[input condition 1],[input condition 2],....,[input condition 8]
$[(x,y)]$,   [1], $[(x+1,y+1)]$, $[(x+1,y)]$,   ..., $[(x,y+1)]$

**input condition
selects $(x+1,y+1)$**

id = (x,y)

FIGURE 9.5
Configuring a *cgp* as a cellular automaton.

mechanism reads the states of its eight immediate neighbors and calculates its new state using the transition function. Because the transition function implements the rules of the cellular automaton and because all of the mechanisms act simultaneously, the *cgp*-v faithfully imitates the actions of the cellular automaton.

We could, of course, add other bits to the id tag for other purposes. For example, to allow several mechanisms at the same coordinate, we could add extra bits to distinguish them.

Billiard Balls and Hot Gases

Once we assign coordinates to a mechanism via the id tag, we can also "move" the mechanism by changing its id tag (see Figure 9.6).

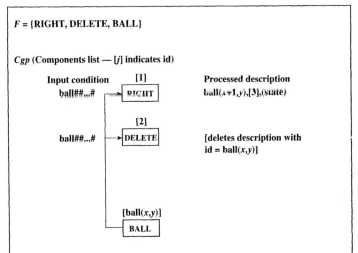

F = {RIGHT, DELETE, BALL}

Cgp (Components list — [*j*] indicates id)

Input condition	[1]	Processed description
ball##...#	RIGHT	ball(x+1,y),[3],(state)
ball##...#	[2] DELETE	[deletes description with id = ball(x,y)]
	[ball(x,y)] BALL	

The net effect of RIGHT and DELETE acting on the description of BALL is to give BALL a new coordinate (x+1,y), effectively moving it one position to the right.

The new description still satisfies the conditions of RIGHT and DELETE, causing the operation to be repeated, so that the ball continues a uniform motion to the right.

FIGURE 9.6
Movement by changing tags.

For example, if the mechanism is located at coordinate (x,y), we move it one step in the x-direction by changing the coordinate part of its id tag to $(x+1, y)$. This technique makes it easy to build *cgp*-v models of mobile particles or agents. Consider a model (such as Echo in Holland, 1995) where simulated organisms (agents) migrate from site to site in a landscape, all interactions taking place within a site. Then the conditions of the mechanisms are set to accept only output from mechanisms with the same (x,y)-coordinate, while migration is achieved by changing the mechanism's id tag and conditions to designate an adjacent coordinate, say $(x+1, y)$.

The textbook model of a hot gas provides another illustration of this use of the *cgp*-v framework. If, for simplicity, we limit the gas to two dimensions, we can identify the gas particles with a set of billiard balls moving on a frictionless billiard table. The simplest version treats the billiard table as a single site, with collisions occurring at random within this site. In effect, we pair billiard balls on the table at random and treat each pairing as a collision.

To mimic this situation within the *cgp*-v, we can allocate a mechanism to each billiard ball. We then assign a unique id tag to each of these mechanisms, all tags having a common prefix that names the site (the billiard ball table). For example, mechanism j at site (x,y) would have the id tag $(x,y)(j)$. Let us also design a mechanism, call it COLLIDE, with a transition function that implements the rules of collision. Further, let the conditions on the input to COLLIDE accept *any* tag that starts with the prefix (x,y). Under the convention for identical id tags, COLLIDE will select, for each of its inputs, a mechanism (billiard ball) with tag prefix (x,y). Thus mechanisms at the site will be paired at random as required by the hot gas model (see Figure 9.7), and their interaction will be provided by the COLLIDE mechanism.

We can carry this example a step further to illustrate how a simple *cgp*-v can generate quite sophisticated behavior. The collision of particle i with particle j results in two inputs to COLLIDE having tags $(x,y)(i)$ and $(x,y)(j)$. Our objective is to use the COLLIDE mechanism to produce updated states for $(x,y)(i)$ and $(x,y)(j)$,

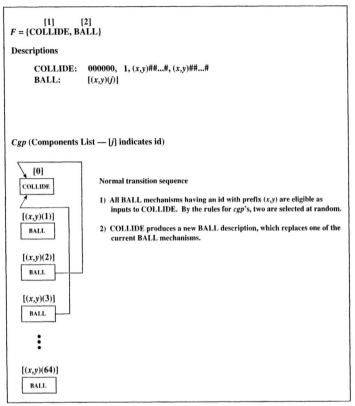

$$
\begin{array}{ll}
\quad [1] \quad\quad [2] \\
F = \{\text{COLLIDE, BALL}\}
\end{array}
$$

Descriptions

   COLLIDE: 000000, 1, (x,y)##...#, (x,y)##...#
   BALL: $[(x,y)(j)]$

Cgp (Components List — $[j]$ indicates id)

   [0]
   COLLIDE

   Normal transition sequence

   1) All BALL mechanisms having an id with prefix (x,y) are eligible as inputs to COLLIDE. By the rules for *cgp*'s, two are selected at random.

   2) COLLIDE produces a new BALL description, which replaces one of the current BALL mechanisms.

   $[(x,y)(1)]$
   BALL

   $[(x,y)(2)]$
   BALL

   $[(x,y)(3)]$
   BALL

   $\vdots$

   $[(x,y)(64)]$
   BALL

FIGURE 9.7
Configuring a *cgp* as a billiard ball model.

states that reflect the effect of the collision. If only one particle were involved, say $(x,y)(i)$, we would simply assign the tag $(x,y)(i)$ to the output of COLLIDE. Because that tag is unique, the output of COLLIDE would replace the original entry on the *csl* with the updated state for $(x,y)(i)$. However, there are two particles, so to do this properly we must assign an interacting *pair* of COLLIDE mechanisms to each collision. The pair of outputs, with tags $(x,y)(i)$ and $(x,y)(j)$, then provide the necessary update.

The example becomes more interesting when we assign a number, which we can think of as energy, to the state of each billiard ball. Let the numbers assigned to particles i and j be u and v, respectively. We set up the transition function of one of the two COLLIDE mechanisms so that it

1. sums the numbers u and v in the state of the two inputs, to form the number $u+v$ (the total energy of the collision);
2. selects a number u' between 0 and $u+v$, $0 \leq u' \leq u+v$;
3. uses this number u' as part of the output state, assigned the id tag $(x,y)(i)$.

Under the convention for replacement on the *csl*, this new state will replace the old entry having the tag $(x,y)(i)$. Effectively, COLLIDE assigns some part u' of the total energy $u+v$ as a new energy for particle i after the collision.

We then arrange it so that the second COLLIDE mechanism in the pair assigns the number $v'=u+v-u'$ to particle j (there are some technical details here that I will not describe). That is, particle j gets the remainder of the total energy. Because $u'+v' = u' + (u+v-u') = u+v$, the total energy of particles i and j is the same before and after the collision. So we have the required conservation of energy in an elastic collision.

The interesting question is, how does this random interchange of energies affect the distribution of energies across the particles (billiard balls) as time goes on? For example, if we start all the particles with the same initial energy, u(initial), it's clear that the random interchange will cause a departure from this uniformity. But what kind of departure?

It can be quickly demonstrated—a few dozen collisions will do—that the energy becomes distributed in a way that has been familiar since the time of Maxwell, the Maxwell-Boltzmann distribution (see Figure 9.8). The energy is distributed about u(initial) so that the number of particles carrying higher energies falls off exponentially. For example, if there are 100 particles with energy $2u$(initial), then the number of particles with energy

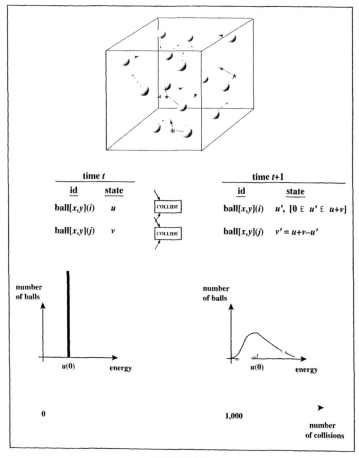

FIGURE 9.8
Generating the Maxwell-Boltzmann distribution.

$4u$(initial) will be about $100^{2u(\text{initial})/4u(\text{initial})} = 100^{1/2} = 10$, the *square root* of 100.

It is a bit surprising that such a subtle distribution emerges from such a simple interaction rule. It adds to our interest that the Maxwell-Boltzmann distribution of energies makes possible catalytic effects. Consider an interaction that requires a total energy $4u$. If we can carry out that interaction in *two* steps, each requiring energy $2u$, the number of candidates for the interaction will be roughly the *square* of the number of candidates available for a single step requiring energy $4u$. Said another way, the reaction rate will be *squared* if the interaction can be carried out in two steps, even though the total energy is the same. We can think of the catalyst as an intermediary that makes the two-step interaction possible.

Biochemical catalysts—enzymes—are the sine qua non of the chemistry of living organisms. So this simple starting point opens the possibility of sophisticated energetics in models of development and evolution like the Echo models (Holland, 1995).

Genetic Algorithms and Cgp-v

As a final example, I'll make a few comments on what it takes to embed a genetic algorithm in a *cgp*-v. An embedding of this kind makes it possible to mimic and study the origin of genetic processes from a less-organized starting point. This particular example does not play a role in the discussion that follows, so the reader not interested in genetic algorithms can safely skip it. For the interested reader without a background in genetic algorithms, a very good introductory work is that of Melanie Mitchell (1996).

Genetic algorithms use operators derived from genetics to effect the evolution of a set, or *population*, of strings. The strings act as chromosomes, typically encoding states or values. They might, for example, be the entries on the current state list in a *cgp*-v.

The genetic algorithm operates in three steps.

1. Selected strings in the population are copied. Selection may depend on some rating of the string, called *fitness* in genetics, or it may depend upon some ability of the mechanism(s) defined by the string to collect enough "resources" to make a copy of the string (*implicit fitness*).

2. The copied strings are then modified, by means of operators suggested by the changes that take place in reproducing chromosomes. Familiar operations like mutation, and less familiar operations like crossover and inversion, are typical.
3. The modified strings are inserted into the population, replacing other strings, so that the population stays the same size.

Earlier I discussed a way to use the mechanisms in a *cgp*-v as a program. That same technique can be used to embed a genetic algorithm in a *cgp*-v. We can assign the elements of the population id tags as if they were stored in registers in a computer. Then, through a combination of special mechanisms like ARITH, NEXT, and INSTR, we can write a "program" that carries out the three steps that define the genetic algorithm.

Even though this procedure is straightforward, it opens up a possibility not exploited to date. What if we back off a step, setting up the apparatus (mechanisms) that makes the genetic algorithm possible, without setting up an actual "program" for the genetic algorithm? If done properly, the *cgp*-v could generate different versions of the genetic algorithm and other string-changing programs. Moreover, via tagging, different versions (and mixes) of these string-changing programs could apply to different populations of strings.

Such a setup makes it possible to observe the emergence of programs that manipulate populations of strings from a precursor system that has no such programs. Persistence plays a key role, as it did in the discussion of Conway's cellular automaton. Different populations, with different versions of string-changing algorithms, would "compete," much as the patterns did in Conway's automaton. Populations with ineffective versions of the string-changing programs would not retain enough coherence to persist and would simply fall apart like an inadequately defined glider in Conway's automaton. Populations that did persist would influence further developments in the *cgp*-v.

A *cgp*-v of this kind, oriented toward the emergence of string-manipulating programs, would let us look into counterparts of a whole range of questions concerning genetics. What kinds of string-changing ("genetic") operators will emerge? How do they serve persistence? Do the same kinds of operators emerge under a wide range of initial conditions? Once we can begin to ask these questions of the models, we can also begin looking to what we know of biological systems to see if counterparts exist. The models can be "tuned" to come closer to extant observations, at the same time suggesting new kinds of observations.

Lessons about Emergence

The principal change in going from a *cgp* with fixed structure to a *cgp*-v is the use of tags and conditions in place of coordinates, along with an enhancement that allows the mechanisms to exploit the tags and conditions. With this modification, mechanisms in the *cgp* can change both the interaction lattice and the mechanisms at the nodes in that lattice. From the examples, we see that we can still construct a fixed structure where that is useful, but the main thrust is a direct approach to mobile agents. This direct approach to mobile agents is important, because earlier studies of complex adaptive systems (Holland, 1995) show that such systems are particularly prone to exhibiting emergent phenomena. It is an arena we will exploit throughout the rest of this book.

Computational Equivalence

In one sense, a *cgp*-v can tell us no more about emergence than a *cgp* with fixed structure. After all, a *cgp* with fixed structure has the power of a general-purpose computer, so we can embed in it any kind of computation-based model. That is, the different versions are equivalent with respect to computational power.

Formalists sometimes treat a search through equivalent structures as irrelevant, or at least as not part of the scientific process. But scientists and mathematicians know the importance of an appropriate framework and diligently search for it—hence the familiar saying, "The problem is nine-tenths solved when the question is asked in the right way." In the present case, equivalence with respect to computational power does *not* mean equivalence with respect to insight. Insights differ even when we confine ourselves to the familiar architecture of commercial computers and something as mundane as word processing. Different programs for word processing provide quite different flexibilities, even when the end product is the same. Such differences are multiplied many times over when we look to broader questions.

The trick is to choose structures that are well-tuned to the question(s) at hand. A study that is straightforward in an appropriate framework can become almost impossible in an inappropriate framework. To take a simple example, normal arithmetic operations, such as multiplication and division, are extremely difficult within the framework provided by roman numerals. Yet the system of roman numerals (augmented by zero) is formally equivalent to the number system we routinely use, where multiplication and division are straightforward. Once we guarantee that the framework has enough computational power, formal equivalence becomes a side issue. The object is to design a framework that makes the operations of interest easy to define and easy to understand.

The appropriateness of the underlying framework is particularly important in the study of emergence because the descriptions and behaviors of systems exhibiting emergence are so diverse. The central question, using the vocabulary we've developed, becomes, what kinds of mechanisms and interactions are *easily* presented and studied within the framework?

Levels of Description

We've already seen that emergent phenomena that are easily understood in one context can be quite obscure in another. There's the closely related question of level of description: laws at one level (say, physics) may completely constrain laws at another level (chemistry), but laws at the latter level can lead directly to answers, whereas a derivation from first principles is tedious or, more likely, infeasible.

Conway's cellular automaton provides a simple example of this interaction of levels. The laws defining the automaton completely constrain the glider pattern, but the macrolaws governing the glider's movement over the cells are what reveal its potential as a signal. The macrolaws provide the insight that enables us to go on to use the glider as a building block for more complex arrangements, ultimately facilitating the proof that Conway's automaton is general purpose, a proof that is almost impossible if one looks only at the defining laws.

This matter of level of description has an extensive history. It was long thought (hoped) that the fifth "axiom of parallel lines" of Euclidean geometry could be proved from the other four. In the nineteenth century, it was shown that one could add a fifth axiom that contradicted Euclid's fifth, while still retaining a consistent axiom system. This discovery led to a whole new range of non-Euclidean geometries, ultimately leading to such concepts as Einstein's theory of relativity.

The point, for present purposes, is that the first four axioms completely constrain what can be achieved by adding further axioms. Indeed, anything that can be accomplished by adding an axiom to Euclid's first four axioms, say adding Euclid's fifth or one of the axioms that contradicts it, can be accomplished within the system of four axioms alone. In the four-axiom system, we can always prove a set of theorems of the form

IF (new axiom) THEN (derivation of theorem based on axiom).

That is, we treat the new axiom as a conditional assumption and carry out derivations based on that assumption. The resulting theorems exactly parallel the theorems that can be derived in the five-axiom system.

Note that we could use other assumptions having nothing to do with parallelism as the IF clause of the theorems just discussed. Indeed, there is an endless range of assumptions that *could* be used. Most would yield theorems that are uninteresting or trivial vis-à-vis questions about geometry. As we've seen repeatedly, the fact that a set of laws or generators fully defines a system does not mean that we can easily derive its consequences. What can be proved or studied in one can also be proved or studied in the other, but the choice makes the difference between a feasible study and one that is only formally possible. Selecting the right alternative among formally equivalent alternatives can make all the difference.

That is really the reason for highlighting carefully selected assumptions as axioms. They define the direction of the study. It required a deep understanding of geometry to formulate an axiom that *both* contradicted Euclid's fifth axiom *and* contributed a set of theorems that enlarged our conception of geometry. Exactly the same point applies in our setting. The trick is to tune the *cgp* to highlight questions about emergence. We need to pick generators and constraints that offer a feasible approach at the right level of description. The insights that lead to these interesting choices often depend on a careful use of metaphor and cross-disciplinary comparisons.

CHAPTER 10

Levels of Description and Reduction

We've seen repeatedly that much complexity can be generated in systems defined by a few well-chosen rules. When we observe emergent phenomena, we ought therefore to try to discover the rules that generate the phenomena. In the particular format we have developed, we need to find a constrained generating procedure that generates the emergent phenomena. In so doing, we will reduce our complex observations of emergence to the interactions of simple mechanisms.

Questions about the capabilities and limitations of constrained generating procedures for carrying out this mission are closely related to the questions that cluster around the philosophical stance called *reductionism*. In its most familiar incarnation, reductionism motivates much of the work in basic science: it holds that all phenomena in the universe are reducible to the laws of physics. Actually, most scientists would make a more cautious statement and say that all phenomena are *constrained* by the laws of physics. Just what is implied by such a view?

First of all, this view does not demand that all explanations be couched *directly* in terms of the laws of physics. It would be both tedious and unenlightening to explain every chemical reaction by using the apparatus and time-scales of quantum mechanics. It is enough to relate various kinds of chemical bonds to quantum mechanical features and use bonds for the rest of the explanation. Even in a model universe where the defining laws are completely known and simple, as in chess or Conway's automaton, much that is observed is determined by large-scale phenomena such as cooperative pawn formations or gliders. Unless we can formulate macrolaws that describe these phenomena, it is difficult to catalog possibilities. Because most large-scale phenomena are emergent, depending on interactions that are more than the sum of local interactions, the difficulties are compounded.

When we can formulate macrolaws that describe the behaviors of emergent phenomena (for instance, the laws of chemical bonding) we gain greatly in comprehension, whether of a model universe or a real one. We can describe a glider in Conway's automaton as moving at a uniform rate in a diagonal direction, in the absence of "interference." Though we know that the behavior of the emergent glider can be reduced to the simple laws defining the automaton, we gain substantial insight into the possibilities in that universe via this macrolaw. Similarly, when we see that masses in the real universe persist in their motion in the absence of "outside" forces, we gain insight into the operations of that universe. Reduction in this case amounts to a conjecture that the complex behaviors we observe can be reduced to a set of simple laws that "define" the universe—the laws of physics. Whether it is Conway's automaton or some real world process, we do *not* expect the emergent phenomena we observe to have simple descriptions in terms of the underlying laws. Indeed, in both cases, we search avidly for simplifying macrolaws.

We can look upon macrolaws as axioms added to the original axioms (the laws defining the model). Typically, the additional axioms will have premises that pick out a range of situations that occur frequently or involve possibilities that lever the system onto new paths. The overall system is still constrained by the original axioms and we could, in principle, derive everything in terms of these original axioms, using the IF (new axiom) THEN (derivation) approach described at the end of the last chapter. As before, there are many possible conditions (macrolaws), and the trick is to pick those that offer possibilities not apparent from direct inspection of original axioms. To reiterate a point made at the end of the last chapter, metaphor and cross-disciplinary comparisons are the key to finding such conditions.

Levels

If we turn reductionism on its head, we add levels to a basic description. More carefully, we add new laws that satisfy the constraints imposed by laws already in place. Moreover, these new laws apply to complex phenomena that are consequences of the original laws; they are at a new level.

We will gain a deeper understanding of emergence if we can deepen our understanding of this concept of levels of definition. The idea of level is intuitive and informal. It is a suggestive idea, but it is subject to ambiguous, sometimes misleading interpretation. If we read "level" wrongly, we can end up with notions of emergence that are so trivial as to destroy the usefulness of the concept ("the levels in organizational charts *show* that organizations are emergent"). Or, we can end up at the other extreme, treating emergence as something holistic that cannot be reduced to anything more basic ("consciousness is distinct from the activity of the central nervous system"). Neither extreme will help us in our quest.

It is here that the precision of the *cgp* setting can be helpful. Because the relation of levels to reduction is a difficult topic, discussions often go astray for lack of clear definition, or because of misinterpretation of verbal definitions. I am about to give a verbal summary of the technical part of this section, but it leaves much to the reader's intuition. An effort to follow the technical notation later in this section should return larger increments in comprehension than a similar effort expended on any other technical section in the book.

Just what is a new level in a *cgp*? The answer turns on one of the basic properties of a *cgp*: the possibility of combining mechanisms to make a more complex mechanism. We started the definition of a *cgp* (in Chapter 7) by characterizing each component mechanism in terms of a set of states S, a set of inputs I, and a transition function f, that specifies the way in which the mechanism's current input and state determine its next state. The *cgp* itself is formed by interconnecting these mechanisms. The objective now is to show that the resulting composite, the resulting *cgp*, can itself be characterized as a mechanism (see Figure 10.1)—a subassembly that can be used to form still more complex mechanisms. When we do this correctly, the transition function of the new mechanism can be reduced to the transition functions of the original mechanisms. We have moved up one level in description.

To show that a composite C is a mechanism, we have to show that there is a single transition function f_C, with corresponding states and inputs, that describes the composite. The way of doing this is actually simple. We combine the states of the component mechanisms of the composite C into a kind of product, where there is a unique product state for each distinct combination of component states. The resulting set of product states, S_C, is the set of states for the composite mechanism. We form a similar product of the inputs of the components and, with some technical manipulations this becomes the set of inputs I_C for the composite. Finally, we define a transition function f_C over these states and inputs, so

Each combination of states for the four components is a state for the composite.

E.g., if the four components are in states 1, 0, 1, 1, respectively, then (1,0,1,1) is a state of the composite.

There are 2^4 such combinations, so the composite has sixteen states.

There are twenty free inputs to the composite.

(If this composite were part of a larger array, some of the free inputs would be connected to the same neighbor.)

FIGURE 10.1

A composite mechanism made from four copies of the primitive mechanism used in Conway's automaton.

that each change of states in the composite is mimicked by a corresponding change in the product states S_C.

Once f_C is defined, we can treat C as if it were a function added to the set of primitive functions F. It can be used as such in the definition of still more complicated cgp's, following the composition procedures just described. More generally, we can produce hierarchical definitions of cgp's, using cgp's defined early in the process as building blocks for later, more complicated cgp's. We gain, thereby, a precise notion of *level*.

Recall that the formal definition of a transition function is given by

$$f: I \times S \rightarrow S,$$

where S is the set of states of the mechanism and $I = I_1 \times I_2 \times \ldots \times I_k$ is the set of input states of the mechanism. We must show, formally, that a combination of interacting mechanisms is a new complex mechanism—a macromechanism—that satisfies the same characterization. That is, we must show that the transition functions of the components act together to define a new transition function. I'll limit consideration to composites that are themselves cgp's because (a) the definition of a cgp provides a precise specification of what it means to have a combination of interacting mechanisms, and (b) we are interested in levels in a cgp.

To begin, let the composite mechanism C consist of n component mechanisms. The first step in defining the transition function f_C of C is to define the states S_C of C. These states, along with the states of the yet-to-be-defined free inputs of C, serve as the arguments of f_C (see Figure 10.1). The definition of the states is easily done. The overall (global) state set of the cgp C is simply the set product (n-tuples) of the state sets of the individual mechanisms,

$$S_C = \Pi_i S_i = S_1 \times S_2 \times \ldots \times S_n.$$

Definition of the input states of C takes a bit more effort. The first step is to divide the inputs of each component mechanism x of C into sets $H_{x,\text{free}}$ and $H_{x,\text{conn}}$, the collections of inputs of x that are free and the inputs that are connected to other mechanisms in C, respectively. To make the notation simpler, reassign the indices to x's inputs so that all inputs in $H_{x,\text{free}}$ have the lowest indices. That is, under the reassignment $H_{x,\text{free}}$ consists of the inputs $\{I_{x,1}, I_{x,2}, \ldots, I_{x,k(x)}\}$ where $k(x)$ is the number of inputs of type $H_{x,\text{free}}$. We can further simplify the notation by forming a single input alphabet $I_{x,\text{free}}$ for x, an alphabet that simply combines values of the individual free inputs. To do this, we form the set product of the $k(x)$ input alphabets in $H_{x,\text{free}}$, proceeding as we did for the states S_i to obtain

$$I_{x,\text{free}} = \Pi_i\, I_i = I_1 \times I_2 \times \ldots \times I_{k(x)}.$$

$I_{x,\text{free}}$ contains all the input values that are assigned from outside C; it acts as the input to the transition function we want to assign to C. We can use the notation

$$I_C = I_{1,\text{free}} \times I_{2,\text{free}} \times \ldots \times I_{n,\text{free}}$$

to designate the values the input can take.

What happens to the inputs in $H_{x,\text{conn}}$? Because these inputs are connected to other mechanisms in C, their values depend on the states of those mechanisms. But those states are components of the global state S_C, so the values $I_{x,\text{conn}}$ of the inputs in $H_{x,\text{conn}}$ are determined by some function g_x of the global state. More formally, there is a function

$$g_x : S_C \to I_{x,\text{conn}}$$

that determines the value of $I_{x,\text{conn}}$ for each global state in S_C, so that

$$I_{x,\text{conn}}(t) = g_x(S_C(t)).$$

When we use this notation, the transition function for mechanism x belonging to C is

$$f_x \cdot I_{x,\text{free}} \times I_{x,\text{conn}} \times S_x \to S_x.$$

But S_x is itself just a component of the global state S_C. So, with some rewriting, we can obtain the function

$$f_x' : I_{x,\text{free}} \times S_C \to S_C.$$

f_x', for given values of $I_{x,\text{free}}$, S_x, and S_C, produces exactly the same value for the S_x component of S_C as f_x does.

With these provisions we can define the global transition function

$$f_C : I_C \times S_C \to S_C,$$

where f_C is defined so that

$$S_C(t+1) = f_C(I_C(t), S_C(t)) = [f_1'\,(I_{1,\text{free}}(t), S_C(t)), \ldots, f_n'\,(I_{n,\text{free}}(t), S_C(t))]$$

for all elements of I_C and S_C. f_C, so defined, completely determines the dynamic behavior of the composite macromechanism; moreover, it has exactly the form required for the transition function of a mechanism.

How to Tile a Cellular Automaton

When we look at a checkerboard, we can revisualize it as a regular arrangement of "clusters" of squares, a process called *tiling* (see Figure 10.2). For example, we might see it as a 4-by-4 arrangement

A 3-by-3 array of mechanisms from Conway's automaton.

A TILE based on the 3-by-3 array.

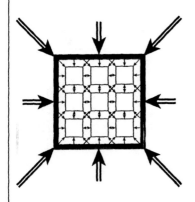

The TILE has 512 states
and
each of its 8 inputs can take
any of 512 values.

The TILEs can be fit together to form an array that looks identical to the original array of Conway primitives. There is a one-to-one correspondence between the behaviors of the two arrays, though the behavior of the tiled automaton appears to be much more complex.

FIGURE 10.2
A tiled automaton.

of square tiles, each of which is made up of a 2-by-2 arrangement of the original squares. We can do the same thing with the geometry of a cellular automaton. We could tile the checkerboard array of Conway's automaton with 3-by-3 tiles, each tile being just large enough to contain a properly centered glider.

Defining a Tiled Automaton

The tiling of a cellular automaton suggests the further step of building a transition function for each of the tiles. We can treat each tile as a composite mechanism made up of the elementary mechanisms (the cells) that form the tile. Then, following the procedures of the previous section, we assign a transition function to each tile.

In the case of the 3-by-3 tiles for Conway's automaton, we would have the following inventory (see Figure 10.2):

1. Each of the nine cells in the 3-by-3 array has two states, so the tile has a total of $2^9 = 512$ composite states.
2. There are eight adjacent 3-by-3 arrays, so the tile has eight inputs. The value of an input at any time is determined by the state of the adjacent array, so the input can take any of 512 values.

The states of the eight inputs to the tile, along with the states of the tile, completely determine the next state of the tile. Thus, we can lay out a table with entries for each possible combination of input states and tile states to determine what the next tile state would be under all conditions. The table would have $2^9 \times 2^9 \times 2^9 \times 2^9 \times 2^9 \times 2^9 \times 2^9 \times 2^9 \times 2^9 = 2^{81}$ entries. That 2^{81} is somewhat larger than 2,000,000,000,000,000,000,000,000. This table—the transition function for the tile—is not only too large for simple pencil-and-paper computation, it is a function that lies beyond the storage capacity of a large digital computer.

The net result is a new cellular automaton made up of the tiles, using the transition function just defined. It is reducible to the original Conway automaton, in the sense that each tile is a composite mechanism made up of the mechanisms (cells) in the Conway automaton. We can rather easily translate any behavior in the tile automaton to a behavior in Conway's automaton, and vice versa.

Reducing a Tiled Automaton

This formal equivalence does *not* mean that corresponding behaviors are equally easy to describe in the two automata. Consider the description of a glider in the case of the tiled automaton. If the glider is centered in the tile, all well and good. We easily pick out the state that corresponds to it. But as the glider moves, its description becomes split between two or more tiles. Moreover, other events can be happening in the parts of the tile not occupied by the glider fragment (see Figure 10.2). These incidentals complicate the state description. Different incidentals yield different states, so there are many states—a very large number, in fact—that indicate the presence of a glider fragment in the tile. The whole description is made still more complicated by the fact that several adjacent tiles may contain glider fragments that do not fit together to make a whole glider.

What if you were presented with the tile automaton with *no* suggestion of its relation to Conway's automaton? You would be faced with trying to find regularities that would help you to understand a transition function with more than 2,000,000,000,000,000,000,000,000 entries! This transition function, taken as fundamental, would provide an extremely complex picture—even more complex than this large number would suggest, because the overall behavior depends on the sequence of interactions between adjacent tiles. After all, we only had to deal with the interaction of two-state cells in Conway's automaton, and that picture was already quite complicated.

What can one do to understand this daunting device? The almost automatic instinct of a scientist is to try to redescribe the states in terms of more elementary components, with the objective of finding components and interactions that are simpler to describe than the original. That objective, more than anything else, motivates scientific reductionism. It has proved a powerful technique for both organizing knowledge and generating predictions in complex situations. Whether we reduce the galactic nebulae in the night sky to their component stars, or a complex molecule to its component atoms, we gain tremendously in our ability to extract regularities and make predictions. In the case of the tiled automaton, the objective is the same.

If we are once again permitted to use our knowledge of the origin of the tiled automaton, the task is simple. We simply reduce the tiles, their states and transition function, to the nine cells from which they were derived. The glider is once more simply described, and many other regularities become obvious. If we must attempt our reduction *without knowing the origin of the tiled automaton*, standard techniques exist for exploring possible decompositions (reductions), but the task is not easy.

Much of the most productive and difficult activity in basic science involves this very task: searching for reductions that uncover previously hidden regularities. The progression over time from observed chemical transformations (such as alchemy) to molecules, atoms, nuclei, quarks, and strings exemplifies a centuries-long effort. The levels of understanding, prediction, and control gained thereby —in areas ranging from medicine to space travel—are familiar elements of everyday life that would not exist without this series of reductions. Unlike the tiled automaton, where the possibility of reduction is guaranteed, the efforts of basic research go forward without such guarantees but encouraged by past successes.

Macrolaws Derived from Conway's Automaton

We've just seen that a macrodescription can obscure phenomena that are readily described and understood at a more elementary level. There are, however, ways of deriving macrodescriptions and macrolaws that extract the essence of an emergent phenomenon. We return to the point that higher-level laws can add to our understanding, even when those laws are fully constrained by more fundamental laws. Conway's automaton, redescribed in the *cgp* framework, lets us illustrate this point in an unambiguous, fully defined setting.

Look again at the glider as a simple example of an emergent phenomenon. The essence of the glider is that it is a changing pattern with a "natural" motion: in the absence of interference, it moves diagonally through the space at the rate of one cell every four time-steps. In this it acts much like a particle moving in physical space according to Newton's laws. In the absence of external forces, the particle persists in its motion with a fixed direction and velocity. It is an interesting test of the *cgp* framework to see if we can extract the glider's natural motion with the framework's help.

Intuitively, the glider is a kind of composite particle, with a shifting configuration, that "floats" through the space of cells provided by Conway's geometry. Can we set up a "migrating mechanism" in a *cgp*-v that captures this intuition?

We saw in Chapter 9 how a mechanism (a billiard ball or a hot gas) could "move" in a space with coordinates. The technique was to change the conditions on the inputs of the mechanism so that each one accepted input from a new coordinate. For example, after the move an input that previously accepted input from the cell at coordinate (x, y), would accept input from the cell at $(x, y-1)$. To apply this technique to the glider, we start by recalling that it has 16 immediate neighbors with which it interacts; these neighbors provide the glider's inputs. If the glider is centered on coordinate (x,y), its inputs are at coordinates $(x, y+2)$, $(x+1, y+2)$, $(x+2, y+2)$, $(x+2, y+1)$, $(x+2, y)$, on around to $(x-1, y+2)$.

If the glider is to move diagonally downward, its inputs change to $(x+1, y+1)$, $(x+2, y+1)$, $(x+3, y+1)$, $(x+3, y)$, $(x+3, y-1)$, on around to $(x, y+1)$. The actual movement, time-step by time-step, is a bit more complicated than this (see Figure 7.4), with the glider moving first straight down, then after a time-step moving straight left, but the technique is the same. As long as the neighbors at the new coordinates remain empty (state 0), the glider continues in this motion. The objective then is to set up a macromechanism that captures this motion, treating interactions with nonempty (state 1) neighbors as a kind of exception to this "natural" behavior (see Figure 10.3).

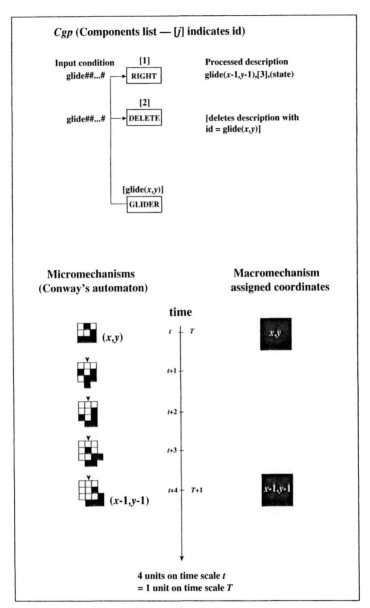

FIGURE 10.3
A macromechanism for the glider.

To accomplish this, we must as always define an appropriate transition function for this "glider mechanism." The transition function must accomplish the following changes:

1. While the neighbors encountered remain empty, the transition function cycles the mechanism through four internal states, corresponding to the four nine-cell patterns the glider cycles through as it moves (see Figures 9.6 and 10.3). Concurrently, the function alters the 16 input conditions to reflect the changing set of neighbors encountered. The output string of the mechanism thus exhibits a repeating set of four internal states, accompanied by changing input conditions reflecting the movement.

2. When a nonempty neighbor is encountered—a kind of interference or "collision"—the transition function must reflect the complicated interactions dictated by Conway's elementary rules. Usually the collision causes the destruction of the glider—its 4-cycle is interrupted—and the simple macrolaw above is no longer of use. At this point, the simplest thing to do is to abandon the macromechanism and return control to a more elementary level.

The abrupt abandonment of a macromechanism "when things go wrong" may seem to undercut our whole plan, but that is not so. What we're encountering in (2) is a change of regime, quite typical of work in science. For instance, a scientist will abandon the macrolaws of chemistry when energies in the nuclear range are encountered. Usually this change in regime involves reduction to a more elementary point of view, often involving a decrease of three or more magnitudes in the dimensions of the system's basic elements (its "cells"). To repeat an earlier comment, changes of three orders of magnitude or more usuually require a new science. In our present simple case, the departure from natural conditions for the glider requires a reduction to the more detailed laws of Conway's "microcells."

The main point really concerns the obverse of reduction. When we observe regularities, we can often move the description "up a level," replacing what may be difficult or even infeasible calculations from first principles. These regularities still satisfy the constraints of the underlying microlaws, but they usually involve additional assumptions. Under these assumptions the regularities persist and a simpler, "derived" dynamics can be used. These additional assumptions are usually described by phrases like "normal" or "natural" conditions. When these conditions are not present, we abandon the macrolevel and return to the microlevel for the more detailed considerations required. Think of the normal conduction of electricity, and the greatly modified, originally unexplainable phenomena when we go to an "abnormal" low-temperature regime (superconductivity). Without the possibility of reduction, behavior under abnormal conditions would simply leave us mystified by a set of unorganized observations.

CHAPTER 11

Metaphor and Innovation

WE STARTED this exploration of emergence by examining two inventions, numbers and board games. These ancient pursuits were contrived long before humans began recording their intellectual achievements. They are simply described, though their discovery was far from simple. In both cases their short, intuitive definitions generated objects that have been fruitfully studied to the present day. Both easily illustrate the "much from little" hallmark of emergence. And both raise the question, how are such inventions produced? Ultimately, to understand emergence, we must understand the process that engenders these inventions. Our quest now enters the broader arena of metaphor and innovation.

In that arena, inventions like numbers and board games epitomize our human ability to reorganize perception through the use of abstraction and induction. Numbers in particular point up the uses of abstraction. To come to the concept of number, almost all details must be dropped from multitudes of observations to arrive at essences like "two-ness," "three-ness," and so on. We do this with the greatest of ease, once taught the trick, but it is no mean feat to discover the trick. It is even more of a feat to recognize the organizing power of numbers. Over the centuries numbers have moved from the counting of herds, to a basis for trade, to Pythagorean and Archimedean theories of the world that replaced myths, to current practice that puts number at the center of the human scientific endeavor (for example, see Newman, 1956).

Board games are not usually accorded the same primacy as numbers, but to my mind they are an equally important cornerstone in the scientific endeavor. In particular, I think board games, as well as numbers, mark a watershed in human perception of the world. A board game qua game only exists because the players act within the agreed-on constraints set by the game's rules. Though the rules must be fully and compactly specified for the game to be "playable," they can be contrived freely relative to the real world, subject only to incidental physical constraints involved in movement of pieces and the like. This freedom from direct physical constraints encourages modifications in the rules, accompanied by empirical judgments as to which rules yield a better game. Each new try amounts to a new miniature universe governed by fully defined laws. It is not a long step from such an outlook to the idea that the world itself may be rule governed.

Above all, board games, unlike numbers in their raw form, capture the dynamic of unfolding actions and their consequences. There is an initial position, and the successive actions of the players give rise to a succession of positions, all within the constraints provided by the rules. Different actions cause different successions. Cause and effect, as well as the possibility of controlling the outcome, become obvious in this context.

The rules of the game expand to become the "rules of logic" as we move forward in time from the board games of the early Egyptian dynasties. Thales' advocacy of "logical speculation"—the counterpart of our search for rules to explain systems exhibiting emergence—moved rule making to a broader interpretation. This logical speculation required adherence to agreed-on rules of reasoning, followed by a comparison of the results with the real world. Thales specifically supported logical speculation as an alternative to traditional myths as a way of understanding nature.

From Thales on attempts to model the world within a logical framework encapsulating cause and effect became increasingly sophisticated. From Euclid's geometry we went on to such triumphs as Kepler's model of the solar system and Newton's laws of the universe. The sine qua non of these models is a small, easily comprehended set of laws that yields a wide range of testable consequences.

In the nineteenth century, Sir Charles Lyell and his compatriots developed their models based on the rate of weathering of mountains and sediment deposition. A whole new conception of the world and its age came into being. Suddenly there was room enough and time for things to evolve, allowing an explanation for the hitherto mysterious skeletons of "monsters" that had been encountered in quarrying. The laws were simple and the conception was testable. Even more important, the laws fit the constraints imposed by Newton's laws. The effect was a cumulative extension of science. Each test of any part of the framework added credibility to the developing whole. When Darwin came along with his astute observations and connections, embedding all within the constraints of Lyell's geology, he gained a credibility that his grandfather, Erasmus, never had, though Erasmus' imaginative insight was of an equal order.

The conjunction of the logical dynamic offered by games with the universal measurability offered by number culminates in a form of modeling that typifies modern science. We see this already in Pythagoras' school based on the relation between number and the musical scale. Number, because of its extreme abstraction, can be attached to almost anything, and the laws of arithmetic nicely reflect various cumulative effects such as the merging of herds, the increase in height as standard blocks are added to a pyramid, the distance traveled at a steady pace, the relation between orbital distance and orbital velocity, the accumulation of sediment under regular weathering, and so on.

Innovation and Creation in the Sciences

I've already expressed my opinion that models are pivotal in the development of science. It's time to enlarge on that view and relate the construction of scientific models to innovation in the sciences.

First, it's important to distinguish the finished product in science from the process that produces that product. The finished product, usually a published scientific paper or book, is presented with careful, step-by-step reasoning, each step following directly and clearly from the previous step (at least for the cognoscenti). The presentation strives for a kind of inevitability, wherein the conclusions are an irrefutable consequence of the starting point. This is, of course, an ideal only approximated in practice, but the best scientific publications are quite convincing in this respect.

This widely accepted scientific standard has given rise to a view, held by some scholars and scientists, that this step-by-step, almost mechanical procedure is the way science is actually conducted. It is a view that marginalizes imagination and creation. But few scientists, if any, actually carry out their research in this fashion. I'll come later to the use of metaphor and aids to the imagination, but let me start with the practice of modeling.

The words of the great theoretical scientist James Clerk Maxwell (1890) provide a good starting point.

> We must therefore discover some method of investigation which allows the mind at every step to lay hold of a clear physical conception, without being committed to any theory founded on the physical science from which that conception is borrowed, so that it is neither drawn aside from the subject in pursuit of analytical subtleties, nor carried beyond the truth by a favourite hypothesis.

He goes on to give a more specific example.

> [Refer] everything to the purely geometrical idea of the motion of an imaginary fluid [that is] merely a collection of imaginary properties which may be employed for establishing certain theorems in pure mathematics in a way more intelligible to many minds and more applicable to physical problems than that in which algebraic symbols alone are used.

Maxwell's other writings make it obvious that his "clear physical conception" turns on the use of devices closely akin to the notion of mechanism developed here. He gives details of the mechanism-oriented fluid mechanical model by which he arrived at his famous equations for electromagnetic fields. Thus, Maxwell moves from a specific mechanical model to the greatest feat of abstraction since Newton formulated his equations for gravitation.

The construction of a model of this kind has much in common with the construction of a metaphor. The *target* model is related to an already constructed *source* model. In the sciences both the source and the target are best characterized as systems rather than as isolated objects. Typically these are systems of interacting (copies of) mechanisms, the kind of mechanisms that underpin constrained generating procedures. For example, Maxwell's papers make it clear that his "collection of imaginary properties" for his "imaginary fluid" is generated by mechanisms described as gears and vortices. This initial scouting expedition, to decide on the mechanisms appropriate to source and target, takes considerable insight and intuition. The result distinguishes innovative science from pedestrian science.

With a target in mind—in Maxwell's case, the large number of electric and magnetic phenomena he was trying to unify—the first step is to select an appropriate source. The immediate instinct in the sciences is to turn to some system that is already well understood. That well-understood source system will have a model specified by a set of laws, of the kind we discussed in the first part of the book. Usually these laws define mechanisms and interactions that generate the phenomena of interest, as in the constrained generating procedures. Decisions about which properties of the source system are central, and which incidental, will have been resolved by careful testing against the world.

As a result of testing and deduction, a well-established model in the sciences accumulates a complicated aura of technique, interpretation, and consequences, much of it unwritten. One physicist will say to another, "This is a conservation of mass problem," and immediately both will have in mind a whole array of knowledge associated with problems modeled in this way. Such use of sources already well tested to gain insight into new problems has much to do with the cumulative nature of the scientific enterprise.

Mary Hesse (1966) makes the point this way:

If some theorist develops a theory in terms of a model . . .
[n]either can he, nor need he, make literally explicit all the associations of the model he is exploiting; other workers in the
field "catch on" to its intended implications—indeed, they
sometimes find the theory unsatisfactory just because some implications the model's originator did not investigate, or even
think of, turn out to be empirically false.

Once the source model is selected, the next task is to carry out a
translation from parts of the source to parts of the target. The
mechanisms of Maxwell's imaginary fluid, his source, must be related to the less-well-understood mechanisms of his target, electromagnetic phenomena. This correspondence between mechanisms
provides a translation that brings the source's aura of technique,
consequences, and interpretation across to the target. The result
must of course be tested in the target domain; if the relationship
has been carefully chosen, years of study and insight are transferred to the target system. Moreover, the new phenomena are related to the older phenomena, perhaps causing reinterpretations
of the older model. This process further enhances the cumulative
enterprise.

Metaphor—A First Look

Once we note that a metaphor also involves a *source* and a *target,* we
can relate this discussion of the use of models for innovation in science to the construction of metaphors. (I should note that I use
the term "metaphor" here in the broad sense, including similes
and other tropes as special cases.) Even a simple metaphor such as
"iceberg, emerald of the sea" involves much more than a source
object, the emerald, and a target object, the iceberg. Both the
source and the target are surrounded by an aura of meaning and
associations. The metaphor causes a kind of recombination of

these auras, enlarging the perceptions associated with both the target and the source. When the metaphor is successful, the interaction produces connections and interpretations that are, variously, surprising, amusing, or exhilarating.

Max Black (1962) enlarges on the point:

> The metaphor works by applying to the principal subject [target] a system of "associated implications" characteristic of the subsidiary subject [source]. . . . These implications usually consist of "commonplaces" about the subsidiary subject, but may, in suitable cases, consist of deviant implications established *ad hoc* by the writer. . . . The metaphor selects, emphasizes, suppresses, and organizes features of the principal subject by implying statements about it that normally apply to the subsidiary subject.

A closer look at "iceberg, emerald of the sea," simple though it is, will help us understand this recombination of auras. The emerald is surrounded by an aura that includes a deep green color, a faceted and brittle hardness, glinting, durability, a connotation of wealth, the romance of the Far East, and so on. The iceberg in sunlight shares some of these properties—glint and a blue-green color, to name two—and it approximates others—several sharply angled, facet-like planes on its visible surface, for instance. The iceberg has its own aura that includes massiveness, melting, danger, the mysteries of the Arctic and Antarctic seas, as well as the properties already mentioned.

It is not easy to spell out how our perception of the two objects is altered by the metaphoric conjunction, because the responses depend partially on individual experiences. For most, the metaphor will activate a train of associations. If you have seen emeralds, but not icebergs, then there is an immediate impression of the iceberg as a thing of beauty. The faceted brittleness might remind you that emeralds can be split by a tap in the right place, suggesting that an iceberg would split similarly under impact. You might even go so far as to relate the juggernaut of wealth, represented by the emerald, to the juggernaut of the seas, represented by the iceberg.

"Iceberg, emerald of the sea" is a simple, unsurprising metaphor, yielding rather simple shifts in perception; but a more subtle metaphor can place the target in a whole new light. "Man is a wolf" is a metaphor that brings up a rich, sometimes jarring, set of associations, from pack behavior to solitary ferocity. Indeed, the associations are so rich that Hermann Hesse could rework them into a major novel, *Steppenwolf*. Many, perhaps most, major novels can be thought of as deeper extended metaphors that cause a profound reconception of the subject matter.

It might seem that we could replace a metaphor with an explicit list of associations, but this is simply impossible. The auras of both source and target are always extensive and amorphous; the conjunction emphasizes some aspects and diminishes others. The metaphor is further tuned by the context in which it appears, which includes both the surrounding subject matter and the observer's experience. Indeed there may be other metaphors, in the observer's mind or in the written context, that shape a given metaphor's meanings. The metaphor catalyzes, and stands in place of, this complicated interaction. Umberto Eco (1994) says it well:

> And this is the office of the supreme Figure of all: Metaphor. If Genius, & therefore Learning, consists in connecting remote Notions & finding Similitude in things dissimilar, then Metaphor, the most acute and farfetched among Tropes, is the only one capable of producing Wonder, which gives birth to Pleasure, as do changes in scene in the theater. And if the Pleasure produced by Figures derives from learning new things without effort & many things in small volume, then Metaphor, setting our mind to flying betwixt one Genus & another, allows us to discern in a single Word more than one Object.

Relations between Metaphor and Model

The following statements seem to apply equally to metaphors and to the *source* → *target* use of models that Maxwell advocates:

1. There is a source system with an established aura of facts and regularities.

   (For models, many of these facts and regularities will be explicitly enumerated, as in a *cgp*, but there will still be the implicit aura of technique, practice, and interpretation.)

2. There is a target system with regularities, and perhaps facts, that are difficult to perceive or interpret.

   (For models, the target may be defined by a collection of observed phenomena that are inadequately explained by the web of existing models.)

3. There is a translation from source to target that suggests a means of transferring inferences for the source into inferences for the target.

   (For models, this translation may be accomplished by matching mechanisms in the source to mechanisms in the target.)

In both cases, the result is innovation; models and metaphors enable us to see new connections. Most of those who are heavily engaged in creative activities, whether in literature or in the sciences, would agree that metaphor and model lie at the center of their activities. A closer look at the ways in which metaphors and models are constructed suggests ways for enhancing the creative process, even though we know little of the mechanisms that underpin the process, and even less of how to direct it. Those suggestions are our next concern.

Cultivating Innovation

Models, particularly source models, are built up from tested components and mechanisms (generators); metaphors point up elements of the source aura that are easily assimilated to the target. In both cases a few well-tried parts can be combined to provide novel insights. Eco (1994) makes the point in the following way:

[As] a follower of Democritus or of Epicurus: he accumulated atoms of concepts and composed them in various guises to make many objects of them. . . . [D]o not playwrights derive improbable and clever events from passages of probable but insipid things, so that they may be satisfied with unexpected hircocervi of action?

Once the building blocks (atoms, parts, generators) are chosen, a large part of the creative act is the selective exploration of the possibilities offered by various combinations. The natural question then is, how do we act selectively?

Selective exploration is closely akin to exploring the possibilities in a game (the game tree). If we think of the rules of a game as its building blocks (generators), then it pays to think again about Samuel's approach to selecting interesting (winning) combinations in checkers. He recasts the game in terms of selected features, where each feature is easily evaluated for different legal configurations. This recasting is closely akin to Maxwell's recasting of electromagnetic phenomena in terms of the more familiar, easily checked properties of his fluid mechanical model. The question immediately arises: How did Maxwell choose his particular form of the source model, and how did Samuel arrive at the features he used?

This intuitive leap is a mystery we cannot fully resolve. No well-defined model of this aspect of the creative process exists, though many, including this author, think that such a model is possible. The lack of a well-defined model leaves the overall process in a shroud of conjecture; still, there are steps we can take to enhance creative activity.

Practice

A part of the answer is familiar to all who strive within a discipline, be it tennis, piano playing, writing poetry, or building a scientific model. The answer lies in the word itself: *discipline.* Only when you are so familiar with the elements (building blocks) of your disci-

pline that you no longer have to think about how they are combined, do you enter the creative phase. If you are a tennis player
and have to concentrate on the elements of each stroke, you will
have little appreciation of the flow of the game—your opponent's
strengths, weaknesses, and strategy. If you play the piano and have
to concentrate on fingering, you will not hear the flow of the music, the "long line." Local concerns drive out global perceptions,
and so it is with other disciplines.

Students wanting to avoid the rigors of discipline will often say,
"I can look that up in the book." That approach works no better
with mental disciplines than it does with muscular skills. No one
would expect to play good tennis or piano through extensive reading without practice, let alone through "looking it up." It is the
same with mental disciplines. The larger view, and effective selection among possibilities, depends on becoming so familiar with
the elements that they no longer require close attention.

Selection of Building Blocks

We cannot bring practice into play until we have decided on the
building blocks that are relevant to the target. If the question or
phenomenon fits readily within an established discipline, the
building blocks are usually given. They have already undergone
testing and selection in the hands of many practitioners. To a large
degree that is what study of a discipline is all about: the acquisition
of the building blocks and associated techniques of that discipline.
If we are studying physics, we learn about motion, mass, and energy, and the forces that transform them; we also learn associated
tools like the differential calculus. If we are studying poetry, we
learn about rhyme, meter, tropes, and standard organizing forms
such as ode, sonnet, or haiku; we also learn critical tools to test our
ideas and inventions. All disciplines have similar requirements.

It is important not to be misled by these deceptively facile descriptions. Many years are spent in acquiring the elements of a discipline, and individual differences come into play in important
ways. Disciplines are like metaphors: they involve complicated
auras that cannot be spelled out in any simple way. "Feeling" for a

discipline only comes from constant immersion. Even this does not guarantee success. Immersion is likely to make one adept, but, as in tennis or piano playing, few make the transition to outstanding. That is part of the mystery that attends our lack of a well-defined model of the creative processes. For some, the elements come together in scintillating patterns, and one moves effortlessly from one "eureka" to another; for others, after equal amounts of immersion, nothing comes easily, and innovations come, if at all, only after great effort; for still others, understanding comes but no innovations whatsoever.

If we understood the creative processes better, we might be able to alter the outcome of immersion, but that is not obvious. It *is* obvious that the bottom-up percolation which provides innovation has a component that lies well below the conscious level. A well-defined model of the process must go beyond "stream of consciousness" descriptions. I said earlier that I do think we can model this component, but we have a long way to go.

What if there is no discipline? What if the phenomenon or question that is of interest does *not* fit within a standard framework? That is part of the problem Maxwell faced. The electric and magnetic phenomena that interested him fit uncomfortably within the scientific framework of his time. It is here that insightful translation—metaphor, if you like—comes into play. Maxwell's familiarity with the various disciplines of physics gave him a source model that served him well in translation.

Again, the immersion was a necessary, but not sufficient, condition for progress. Without the immersion, the source model would not have been available to Maxwell. We know little of the insight that enabled Maxwell to pick the source model he did. It involved something about seeing the commonality of the underlying mechanisms, but that is at best part of the picture. The aphorism "Look for the underlying mechanisms" is helpful but vague. We only know that having a familiarity with several "nearby" disciplines, when the target does not fit well within an established discipline, will enhance the possibility of a source → target transfer. The perception of what's "nearby" is a part of that still mysterious trait we call insight.

Exploring Larger Patterns

Once use of the selected building blocks is habitual, the principal step is that of perceiving seminal patterns in the combinations generated by the building blocks. Because the discovery of a new building block is a rare event, most innovations stem from the generation of new combinations of well-tried blocks. The internal combustion engine and the digital computer, for instance, were first constructed of building blocks that had been familiar for decades, or even centuries (gears, Venturi's aspirator, Galvani's sparking device, and so on, in the case of engines; Geiger's particle counter, cathode ray tubes, and so on, in the case of computers—see Burke, 1978).

Even when many people share the same building blocks, different innovations can arise. The standard building blocks provided by words in a language, or themes in music, yield a never-ending flow of innovations. Of course, the disciplines of writing and music involve other higher-level building blocks, constructed from these elements. However, the higher-level building blocks are still patterns of words and chords, much as the words and chords themselves are generated from still more primitive elements, alphabets and notes.

We see again the creative obverse of reduction. A new word uses the same old alphabet, bringing across familiar elements of pronunciation and easy recognition, but the word will have its own aura and niche. Often the word will be a combination of (parts of) already familiar words so that, like a metaphor or Maxwell's source → target use of models, the new word acquires parts of its aura from extant forms. As with the reduction hierarchies in science, the lower levels supply constraints that must be satisfied by the higher levels. But the higher levels involve regularities that are best understood by laws stated in terms of the building blocks of that higher level. The new word, once inserted in the dictionary, obeys rules that apply to all words in the dictionary.

Unfortunately, we know little about the processes that underlie the perception of larger patterns and regularities in the generated combinations, even though that step is critical to innovation. Indeed, we know little about the more overt process of *visual* pattern perception, despite years of effort in psychology and artificial intelligence (AI). In part this is because few psychological models of vision, and none of the standard models in AI, are saccade based (see Chapter 5). Yet experiments have made it clear that saccades play a key role in human perception. If you are asked a question about a picture or scene, your saccadic eye movements are far from random. The jumps land on relevant parts of the scene and focus attention there. It is much like determining the features in a checkers configuration, à la Samuel (1959), that are relevant to play of the game at that point. Our ignorance of the mechanisms that pick out salient features in visual pattern recognition leaves us with little to go on when it comes to the more subtle problem of detecting patterns in generated combinations.

The earlier discussion of neural nets with cycles does offer a few clues, at the microlevel, about the internalization of salient features in repeated situations (via cell assemblies and phase sequences). Cross-inhibition and fatigue provide the anticipation that can guide saccades. If the anticipation is borne out regularly, the relevant features (cell assemblies) are established as part of the repertoire. But there is a substantial distance between mechanisms at this level and the perception of patterns in generated combinations. About all that comes across at this higher level is the relevance of exposure and practice.

When we play a game regularly, we begin to see certain kinds of patterns—in chess, they go by names such as "fork," "pin," and "discovered attack"—that serve as building blocks for longer-term, more subtle strategies. We use these patterns to select parts of the game tree, ignoring the overwhelming range of nonproductive or uninteresting possibilities that make up most of the tree. It is this ability that distinguishes the human approach to chess from the programs used in the current highly publicized human-computer chess tournaments. These programs are largely committed to an exhaustive search of possibilities, which requires tremendous increases in computing power for each additional level of lookahead (see Chapter 3). In contrast, humans who are good at board games look at relatively few combinations, avoiding vast nonproductive "deserts" in the rest of the game tree.

Certainly the human approach may miss the optimal line of play. However, the rampant bushiness of the game tree assures that an exhaustive search can only look so far ahead; thus it too must fail in finding the optimal line of play in all but the most simple games. The same argument holds for the broader realm of generated (law-defined) systems, characterized here as *cgp*'s. Constrained by feasibility, both human and program must find promising sequences in a vast space of combinations. Anticipation-based salient features and sequences, in the manner of directed saccades, offer great advantages over exhaustive search— the more so because the object is improvement, not optimality.

The history of science, in each of the disciplines, shows a progression of theories that explain a wider and wider range of phenomena. Whatever the long-range prognosis, the complexity of the universe makes arrival at an "optimal" theory unlikely, at any level, in the foreseeable future. Even for a universe as simply described as chess, after centuries of intense study we have no optimal theory. How likely is it that we will attain that goal for a real universe, for which even the rules are unknown?

That improbability is actually not a very severe constraint if we are *not* searching for optimality. When we face complex situations, our objective is almost always to "do it better." That "better," whether model or metaphor, may be something new and unexpected or it may be an improvement on something familiar. Over the centuries our level of competence in chess has continually increased, as evidenced by levels of play in recorded games, without exhausting the possibilities of the game. Over those same centuries our models (theories) in science have built in cumulative fashion, continually increasing in scope and prediction, and again the end is nowhere in sight. In the realm of metaphor, literature, and poetry, the cumulative effect of doing it better is less apparent, but new insights keep appearing. Not even the most dedicated advocate of canon would talk of a "best" poem, play, or novel.

From this point of view, the central question in the study of cognition would seem to be the nature of the mechanisms, and interactions, that underlie our ability to discover larger patterns in generated processes. Our search seems to be steered, as in the case of Copycat (Mitchell, 1993), by the auras that surround the relevant building blocks. The metaphors, analogies, and models that result guide us through the complicated mazes presented by the world. We will not understand cognition until we can model this process.

In Brief

There is no simple crank to turn for innovation, such as going round and round between more facts ("the book") and incremental revisions of hypothesis. It *is* possible, on occasion, to follow this simple "hypothesize, test, and revise" pattern, but innovation requires more: at the start one must select a target. This usually happens naturally, even serendipitously, because of some unexplained phenomenon or some question posed. Then, framed in terms of our general setting, come two major steps: (a) discovery of relevant building blocks, and (b) construction of coherent, relevant combinations of those building blocks.

In the sciences, we've already noted, new building blocks are usually constructed by combination at a more detailed level: proteins from amino acids, amino acids from atoms, atoms from nuclear particles and electrons, and so on. This seems less frequently the case in literature and the arts, a point to which I'll return. In all cases a new building block opens whole realms of possibility, because of the new prospects for combination with extant building blocks.

Once a set of building blocks has been chosen, innovation depends on selection from among the plethora of potential combinations. The possibilities are so numerous that the same building blocks can be used over and over again without seriously impairing the chances for original discoveries. Think of the standard building blocks provided by words in a language, or folk themes in music. The key to handling this complexity is the discovery of salient patterns in the tree of combinations. Creative individuals exhibit a talent for such selection, but the mechanisms they employ are largely unknown.

It is a matter of speculation, but worth examining, that the mechanisms of selection in the creative process are akin to those of evolutionary selection, simply running on a much faster timescale. Even something as simple as a speedup can revise a model to the point of changing our outlook. A lapse-frame movie of a wild grapevine moving up a tree looks remarkably purposeful. A lapse-frame animation of geological evolution shows the fluid, responsive, coherent movement of clouds in the sky. A lapse-frame animation of the evolution of some family of organisms would almost certainly show the tentative probes, withdrawals, redirections, and cumulative construction we associate with creative activity. In both evolutionary and creative exploration we encounter patterns and lines of development (strategies) that emerge under selection. And in both cases emergent building blocks propagate their effects in cumulative ways, through recombination and interaction. We've already seen several examples of this propagation in the earlier discussion of reduction (Chapter 10), and more detailed discussions of the mechanisms are given in *Hidden Order* (Holland, 1995). There is much to be learned, I think, by modeling cognition via a translation of the mechanisms of natural selection, mimicking Maxwell's translation from gears to fields.

Poetry and Physics

When we look at the creative process, it's interesting to compare the two great *P*'s of human intellectual endeavor, Poetry and Physics. Each produces deep insights into the world that surrounds us, but their disciplines seem *very* different. However, this dissimilarity makes even a brief comparison useful in enlarging our understanding of emergence and the creative process.

Despite the differences, some of them deep, creative activities in poetry and physics have much in common. Both the poet and the physicist strive to get beneath the surface of events, the poet concentrating on the human condition, the physicist on the material world. Both depend on the guidelines and tools that come from tutoring, discipline, and experience, working within forms and constraints suggested by their respective disciplines. For the poet, discipline supplies format (the sonnet), universal myths (the legend of Orpheus), and symbols (the rose); for the physicist, discipline supplies standard models (the billiard ball model), universal laws (the conservation of energy), and mathematical formalism (differential equations). For both, broken symmetries and rhythmic shapes signal possibilities and opportunities. For the poet, a broken rhyme invokes close attention and unusual interpretation; for the physicist, a lack of symmetry in an interaction suggests new particles. Intuition, taste, and leaps of faith based on experience are indispensable to the production of either a poem or a scientific theory. Even if one wants to break the canon, discipline plays a critical role; both poet and physicist subscribe to Bacon's aphorism, "Truth comes out of error more readily than out of confusion."

The differences in product are as instructive as the similarities in process. The poem aims at obliqueness and ambiguity to engage the reader at multiple levels; a scientific theory aims at elimination of ambiguity through a rigorous line moving from premises to conclusion by truth-preserving steps. The poet relies on the conventions of grammar to tie familiar elements into a framework within which to present the unusual. The ambiguity of language is exploited within this framework to direct the reader to levels of meaning not obvious on the surface. The scientist relies on the conventions of logic and mathematics to tie observations into a framework that makes prediction possible. The generalizations provided by mathematical characterization direct the practitioner from specific observations to widely applicable laws.

In a sense, the poetic framework is too loose, whereas the scientific framework is too tight. The looseness of the poetic framework limits the possibilities for a cumulative structure. Though the discipline of poetry has evolved, particularly in the accumulation of technique, the insights remain much the same. Aristophanes' plays hold their own in the modern context. In scientific theory, the rigorous use of prior models as sources for newer, more encompassing models provides a regular succession. Kepler's insights have been succeeded and surpassed by Newton's insights, which in turn have been succeeded and surpassed by those of Einstein, and so it is likely to continue beyond the foreseeable future. Yet this very rigor restricts the scientist's ability to deal with the broad, ill-defined domains that are so much a part of human experience—domains characterized by words like "beauty," "justice," "purpose," and "meaning." The insights of poetry far surpass those of science in these domains.

It is not impossible that poetry and physics can be brought into closer conjunction. Hermann Hesse's *Das Glasperlenspiel* is suggestive. Perhaps there is a "game" with the rigor of a *cgp* that permits insightful combinations of the powerful symbols of poetry. It is a vision that has held me since the days when I first read Hesse's masterpiece.

CHAPTER 12

Closing

THE WORD "closing" has two senses, and both apply here. The first meaning is "ending" or "finishing"; the second is "coming closer."

In closing (first sense), I will summarize some of the main points of this book. The summary involves some of the difficulties that accompany an attempt to summarize a poem, even though there is little of the poetic in this presentation. Much of the meaning is bound up in the way the presentation unfolds. Moreover, at this stage emergence is a subject as ill defined as "purpose" or "justice." The uncertainty of definition forces us to rely on partial descriptions, which in turn rely heavily on context. As a consequence, this summary mainly suggests loci for increased attention.

In closing (second sense), I emphasize that we're nowhere near the end of the exploration of emergence. We're not even at the end of my own thoughts about emergence, but I'd have to employ a more intricate formal apparatus to express them. Still, we have made progress. Though constrained generating procedures supply neither necessary nor sufficient conditions for emergence, they do capture many of the important elements of emergence. At any rate, they constitute a way station. This part of the closing seeks to convey some thoughts about where we can go from that way station.

Closing as a Summing Up

Two cautions: (a) This recapitulation extracts its points from the context that supplies much of their meaning. Meaning is less dependent on contextual remarks when there is an overarching theory, but our exploration has not yet progressed that far. Without that theory, the key ideas depend for their substance on the aura of surrounding remarks. The summary that follows should be treated more as a set of reminders of relevant criteria than as a self-sufficient summary; (b) I am not attempting to list a set of necessary or sufficient criteria for emergence, not even in an informal sense. There are certainly examples of emergent phenomena where one or more of these criteria do not apply—so the criteria are not *necessary* in the formal sense. It is also likely that there are examples of systems where the criteria are met but no emergent phenomena ensue—so the criteria are not *sufficient.* I *am* claiming that you are much more likely to encounter emergence, in models and in the world, if you look at processes or systems exhibiting these criteria.

Basic Concepts

While there is, as yet, no comprehensive formal framework for the study of emergence, well-defined technical concepts serve as foundation blocks for such a framework. The most important concepts encountered along the way come in three clusters: (a) purely mathematical concepts, (b) system concepts, and (c) general, informal concepts. Though I will not repeat the definitions given earlier, I will give short commentaries suggesting the role of each in the exploration.

First, there are the purely mathematical (logical) concepts.

■ *Equivalence class.* The formal counterpart of cropping details in order to emphasize selected features. The equivalence class consists of all objects in the area being studied that share the selected features.
■ *Function (mathematical map).* The formal counterpart of a list of correspondences, where each element in one list (the domain set) is assigned a corresponding element in the other list (the range set). This elementary concept is at the core of

the central concept of system theory, the *transition function* (see below).

■ *Set of generators.* Used here to arrive at a formal counterpart of rules or laws. We are given an initial set of elements (corresponding to pieces in a game) and legal ways of combining them (corresponding to the legal configurations specified by the rules of the game). This concept underpins both a major theme in this book, modeling, and the general setting used to unify this discussion of emergence, the constrained generating procedures (*cgp*'s).

■ *IF [ ] THEN [ ] clauses.* Used here as a way of specifying allowable interactions, particularly between agents (see below). Stimulus-response actions—IF [stimulus] THEN [response]—provide the simplest examples of such usage. IF [ ] THEN [ ] clauses are the heart of the flexible, conditional responses that give digital computers their tremendous power.

Next, there are the concepts more closely tied to systems and games.

■ *State.* The state of a system treats as equivalent all past histories that offer the same future options; accordingly, one need only know the current state to determine possible future states. For board games there is a direct translation from the arrangement of pieces on the board to the state of the game; the state of a more complex system—say, a neural network or a system in physics—may not be so easily defined.

■ *Transition function.* The transition function takes as its arguments (domain set) every legal pairing of states and inputs (input signals, applied forces, or the like). For each such pair it specifies the state that will result next. (The definition of a transition function can be extended so that the next state is determined probabilistically, rather than deterministically, but that has not been an urgent concern here.) The tree of possible move sequences in a game (the game tree) supplies a simple example of a transition function; it unifies the study of games as apparently distinct as chess and poker.

- *Strategy.* Strategies arise when a system has inputs that influence its state sequence and the states are ranked (say, according to desirability). A strategy is specified when, for every state in the domain set of the transition function, a particular input value is specified (as when each opponent in a game chooses a distinct move). That is, a strategy is specified by a function mapping states to the set of input values. If the system has several inputs (for instance, one input for each player in a game), and a strategy is selected for each input, then a unique sequence of states (a trajectory) is determined from any starting state. In the game of chess, for example, if the two players have fully determined strategies, the outcome of the game is determined. (In practice, it is rare to encounter fully determined strategies in systems that are complex because there are so many options, but the concept helps to unify a diverse range of ideas about games and systems.)

Finally, some very general concepts have played a key role at every stage of the development.

- *Building block.* Much that is bound up in the concept of a building block is captured by the technical definition of a generator, but the aura of meaning that surrounds this concept is considerably larger. Building blocks range from mechanisms in physics to the way we parse the environment into familiar objects; they provide a way of extracting repeatable features from the perpetual novelty that attends systems exhibiting emergence.
- *Model.* This is the concept most important to this study of emergence. The critical steps in constructing a model are selection of salient features (equivalence classes) and laws (generators and transition functions) governing the model's behavior. These steps are guided by metaphor and source → target model transfer. Emergence and innovation cannot be understood without a thorough understanding of models.
- *Agent.* Most systems that exhibit emergence can be modeled

in terms of the interaction of agents. Agents, which can range from "billiard balls" in a random interaction model to organisms that adapt and learn, offer the quickest route to building models that exhibit emergence. Agents have not received top billing in the present discussion only because they were the central concern of *Hidden Order* (Holland, 1995).

Recapitulation

I offer the following recapitulation, based on these concepts.

1. *Emergence occurs in systems that are generated.* The systems are composed of copies of a relatively small number of components that obey simple laws. Typically these copies are interconnected to form an array (checkerboards, networks, points in physical space) that may change over time under control of the transition function.

2. *The whole is more than the sum of the parts in these generated systems.* The interactions between the parts are nonlinear, so the overall behavior *cannot* be obtained by summing the behaviors of the isolated components. Said another way, there are regularities in system behavior that are not revealed by direct inspection of the laws satisfied by the components. These regularities both explain (parts of) the system's behavior and make possible activities, and controls, that are highly unlikely otherwise (as when a strategy, based on certain pawn structures, enables a player to win consistently at chess).

   The definition of the generated system, though it determines all the rest, is no more than a simply described starting point; subsequent activities can be determined only by extended examination and experiment. In this sense, more comes out than was put in.

3. *Emergent phenomena in generated systems are, typically, persistent patterns with changing components.* Emergent phenomena recall the standing wave that forms in front of a rock in a fast-moving

stream, where the water particles are constantly changing though the pattern persists; in this they differ from concrete entities, such as rocks or buildings, that consist of fixed components. The pattern of a moving, changing pawn formation, or the reverberations in a set of neurons, are cases in point. Organisms are also persistent patterns; they turn over *all* their constituent atoms in something less than a two-year span, and a large fraction of their constituents turn over in a matter of weeks.

Only persistent patterns will have a directly traceable influence on future configurations in generated systems. The rules of the system, of course, assure causal relation among all configurations that occur, but the persistent patterns are the only ones that lend themselves to a consistent observable ontogeny.

4. *The context in which a persistent emergent pattern is embedded determines its function.* Because of the nonlinear interactions, a kind of aura is imposed by the context. The different uses that can be made of a *glider,* in interaction with other patterns in Conway's automaton, provides a simple, fully defined example. Multifunctionality in biological systems provides more sophisticated examples. For instance, a set of three bones that begins by providing a flexible linkage in the gill arch of a fish becomes, in a later incarnation, a linkage allowing the extra-wide extension of a reptile's jaw and, still later, a linkage in the inner ear of a mammal. The three-bone linkage is well defined and is preserved over time, but its function varies according to context. It is this changing aura that makes it hard to characterize emergent phenomena a posteriori.

5. *Interactions between persistent patterns add constraints and checks that provide increasing "competence" as the number of such patterns increases.* As a simple example, consider the way in which redundancy in the DNA code facilitates correction of local errors in the duplication process. The emergent competence of an ant colony or a neural network, as the number of individuals increases, provides a more sophisticated example.

Nonlinear interactions, and the context provided by other patterns (sometimes simply copies of a given pattern), both increase this competence. In particular, the number of possible interactions, and hence the possible sophistication of response, rises extremely rapidly (factorially) with the number of interactants.

6. *Persistent patterns often satisfy macrolaws.* When a macrolaw can be formulated, the behavior of the whole pattern can be described without recourse to the microlaws (generators and constraints) that determine the behavior of its components. Macrolaws are typically simple relative to the behavioral details of the component elements. The law describing the behavior of a glider in Conway's model universe is a clear example.

7. *Differential persistence is a typical consequence of the laws that generate emergent phenomena.* In Samuel's checkersplayer, for example, new strategies (new weightings) come from revisions of weightings of strategies that have been persistently successful against opponents. In neural networks, it is the persistent reverberatory patterns that become the elements—Hebb's cell assemblies—of more sophisticated behaviors. And, in Darwinian evolution, the patterns that persist long enough to collect resources and produce copies are the ones that generate new variants.

Differential persistence takes a variety of forms. Some patterns persist only as long as they do not encounter other patterns. Others persist through some interactions, while undergoing dissolution or transformation in others. Still other persistent patterns interact with only a few other patterns, simply maintaining their form in all other contexts.

This differential persistence can have strong effects on the generation procedure. The patterns that are likely to take a significant role early in the generation process are those that persist through many kinds of interaction. Then, many possible combinations are sampled, increasing the likelihood that some more complex persistent pattern will be discovered. These *gen-*

eralist patterns can also provide a niche for *specialist* patterns that have a more restricted range of interaction. Occasionally, a specialist will fit with a generalist in a kind of symbiotic way, with the specialist protecting the generalist from interactions that would cause its dissolution.

The formation of a *default hierarchy* (see Figure 12.1) illustrates this kind of interaction. Think of a simple organism, say an ant,

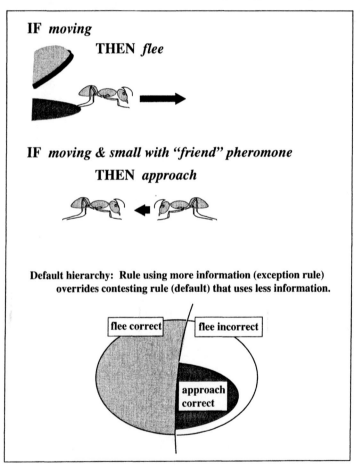

FIGURE 12.1
A default hierarchy.

and assume that it is guided by a general rule that causes it to flee whenever it detects a moving object. This rule will serve the ant well in many contexts, because most large moving objects in its environment can cause its "dissolution." The rule will be tested frequently, often saving the ant from devastation and never causing direct damage, even when the action is inappropriate. On the other hand, this same rule will cause the ant to avoid other moving ants, with long-term deleterious effects. There is clearly a place for a second more special rule that corrects the first rule: IF the object is moving *and* small *and* exudes the "friend" pheromone, THEN approach the object. If this specialist rule takes over whenever these more particular conditions occur, a symbiotic relation results. The specialist prevents the generalist from "making mistakes" that could cause long-term damage to the whole, while the generalist acts to prevent dissolution in a wide range of situations that would not invoke a response from the specialist.

Generalists are much better tested for persistence than specialists, because the generalists receive many trials in a wide range of situations, in the same time that the specialists receive just a few trials. As a result, the generalists supply a firm niche for the specialists that can interact with them. The specialists, being less frequently tested, are much more subject to dissolution in an "open" environment. This relation of sampling rates enters into the formation of macrolaws and layered generation procedures. *Hidden Order* (Holland, 1995) has much more to say about default hierarchies.

8. *Higher-level generating procedures can result from enhanced persistence.* Cross-supporting interactions (for example, symbiosis and Eigen and Winkler's 1981 hypercycles) often provide enhanced persistence for the component patterns. When these patterns with enhanced persistence satisfy simple macrolaws, a new generating procedure is overlaid on the original.

A generating procedure so produced still obeys the laws of the underlying generating procedure, but it yields patterns extremely unlikely on an a priori inspection of the original generator. Higher-level generating procedures can be greatly amplified by enhanced persistence, completely "taking over" the underlying procedure.

Darwin's discussion of the origin of the mammalian eye provides a fine example of an induced higher-level generating procedure, a procedure that makes likely a pattern which would be quite unlikely on an a priori inspection of the ur-elements. Prior to Darwin, it was argued that something as exquisitely organized as the eye could only have been produced by a *designer*—it could not have been produced by a chance assembly of parts. Indeed, eyes, and most biological objects, are extremely unlikely if one looks to their likelihood only as a random selection from the vast number of objects that can be formed from atoms.

Darwin's macroargument is nowadays bolstered by the deeper layer provided by the molecular biology of the eye, a set of factors unknown at the time he was writing. We now understand that the energy of light alters the bonds of certain relatively simple biomolecules, setting off a chain reaction that can, for instance, cause neurons to fire. Light-sensitive compounds, lenselike crystalline compounds, neurons, and so on, serve as building blocks (generators) for the higher-level generating procedure that eventually produces eyes. Darwin's step-by-step procedure for the origin of eyes can be recast in terms of this overlaid generating procedure. What was extremely unlikely on inspection of the generating procedure based on the interaction of atoms to form molecules, becomes likely—almost inevitable—once we take into account the formation of a higher-level generating procedure.

The eye-generation process has been repeated at least twice in evolution, in mammals and in cephalopods, using different building blocks (compounds, cell morphologies, and so forth) to achieve the complex design of an eye with the familiar parts (lens, adjustable focus, retina). In some ways the eyes of cephalopods (like the squid and the octopus) are even better designed than those of mammals. It is evident, then, that once the building blocks are taken into account, eyes are not so unlikely.

This transformation from the extremely unlikely to the likely is a major characteristic of systems exhibiting emergent phenomena. Even when the simplest persistent patterns are infrequent in a generating procedure, they will eventually occur if the system runs for any length of time. Once they occur, they will by definition persist, making them candidates for combination with other persistent patterns (other copies or variants). At this point larger patterns with enhanced persistence and competence can occur. Once some initial building blocks are discovered—simple membranes, the Krebs cycle, differential adhesion of components, and so on—the number of combinations yielding viable organizations goes up dramatically. The usual argument that evolution requires long sequences of improbable discoveries, and so is "slow," misses this point. The unlikely will become likely if one allows for a layered series of generating procedures.

Closing as a Way Station

Why is it that this expedition has come to a way station rather than the destination? What stands in the way of a scientific understanding of emergence?

It is possible, at this point, to discern some of the obstacles between our present position and a better understanding of emergence, and I will describe them here. There is one larger issue, however, that I will avoid. It may be that the parts of the universe that we can understand in a scientific sense—the parts of the universe that we describe via laws (axioms, equations)—constitute a small fragment of the whole. If that is so, then there may be aspects of emergence we *cannot* understand scientifically. Nevertheless, we already know that there are lawful fragments in which we can observe and explain emergence. It is those fragments with which I am concerned.

Mathematical Obstacles

At the formal level, we have a remarkable small body of mathematics that deals directly with nonlinearity. Almost all of the well-established tools of mathematics—partial differential equations, probability theory, Markov processes, and so on—are built upon assumptions of linearity and additivity. Even those parts that take nonlinearity as a subject matter usually depend upon linear approximations. As a result, most scientific models, at one level or another, are based on assumptions of additivity and linearity.

The one area where we deal constructively and regularly with nonlinearity is the construction of computer-based models and simulations. Almost no theoretical guidelines exist for that endeavor. Techniques are learned by apprenticeship. For instance, when we construct computer-based models, we usually translate extant models based on partial differential equations into computer programs. The resulting models carry along the linear underpinnings, rather than exploiting the possibilities for rigorous nonlinear models that the computer makes possible. Cellular automata constitute a notable and interesting exception, but experiment still rules after forty years, and the available theory is just beginning to penetrate beyond the obvious.

It is worthwhile to emphasize again that computer-based models offer a halfway house between theory and experiment. They have the rigor of mathematical models without the generality, while allowing the selection and repeatability of good critical experiments without the enforced connection to reality. We have barely begun to exploit the possibilities of nonlinear modeling in this realm, but computer-based models will certainly improve our understanding of emergence.

Cognitive Obstacles

At a deeper level, our abysmal ignorance of most aspects of cognition presents a serious deterrent to the understanding of emergence. Even at the most elementary level, our models of primate vision are woefully inadequate. As a consequence, our understanding of perception barely exceeds the superficial. We model pattern recognition as if the whole scene were presented to the brain in one large, static array of pixels (bitlike picture elements). Reality for primate vision, as I mentioned briefly in the discussion of neural networks in Chapter 5, is mediated by a series of saccade-based, snapshotlike inputs, each of which captures only a local excerpt of the scene. We know from experiment that saccades are partially directed by higher activities in the central nervous system, and they involve local anticipations. Clearly, saccades have much to do with the selection of salient features and this process, in turn, has much to do with the way we model the world.

If we cannot model salience and pattern recognition at this level, how much more difficult must it be to model the kinds of selective attention that go into metaphor making and conscious model building. We have seen that the aura surrounding the source and target in metaphor making and conscious model building has much to do with their productiveness in innovation. Are there counterparts of saccadelike actions in our central nervous system, mediating selection and anticipation at that level? Such processes would seem well suited to Hebb's cell-assembly-oriented view of the operation of the CNS. However that may be, we do not yet have a firm understanding of the way the central nervous system selects relevant pieces from the never-ending, perpetually novel torrent of sensory information it receives. Until we do, our understanding of the aspects of emergence related to model building, metaphor, and innovation will languish.

A Mental Obstacle: "The End of Science"

In recounting the foregoing obstacles I am setting forth an agenda, not bemoaning an end to our efforts. Because so many of the problems that baffle us—ranging from the control of economies to understanding consciousness—involve emergent phenomena in a crucial way, one might casually infer that this fact somehow signals an impassable barrier. In such a view, we've gone as far as science can take us.

It has again become fashionable in some quarters to talk about the end of science, holding that problems like emergence will lie forever beyond scientific ken. This talk recalls the pronouncements of some senior scientists at the end of the last century. Looking even further back, similar assertions have been made at the end of each century since the time of Newton. These scientists offer the view that the main work has been done and that all the rest is a matter of filling in details. It is almost inevitable that each generation will think it has laid the foundations for all future work. It is even true. Were it not so, science would not be the cumulative enterprise that it is. But that does not prevent extensions, and deepenings, of the foundation. Despite the pronouncements, science continues to move on to an ever broader perspective.

It is now the end of a century that has seen vast foundational changes, and it is the end of a millennium. One might reasonably expect strong, end-of-the-millennium proclamations from some of the scientists who have contributed to these foundational changes. Yet they are not forthcoming. The current talk originates largely from people who are not scientists. Their talk sets forth the view that certain subjects deemed to lie within the purview of science in fact lie forever beyond its grasp. This declaration differs in form from most past utterances, and it is not so well informed.

There is little evidence to support this end-of-the-millennium view, and large amounts of evidence to the contrary. There are still many "big" problems, ranging from "a unification of theories in physics" through "the origin of life" to "the nature of consciousness," and they are steadily yielding ground to scientific inquiry. The very scientists who have contributed to the deep foundational changes of the twentieth century (those still living) see these horizons, and many are actively engaged in constructing the roads that will take us there. I know of no major living scientist that feels "all the rest is detail." Most of all, there is no lack of progress, in the accepted scientific mode, in understanding these problems. They may keep us occupied, at a deep level, for a very long time, but that does not constitute evidence of lack of progress.

The history of science bolsters this point. More than two millennia elapsed between Greek observations of static electricity (thought to be a property of amber) and Maxwell's theory. Even after this long, sometimes sporadic, period of study, we continue to enlarge our *fundamental* understanding of electromagnetic phenomena, while steadily enlarging the range of applications based on that understanding. Today our understanding of electromagnetic phenomena is more fundamental than it was even a couple of decades ago, let alone a century ago. So it is with the current great problems. On the evidence, we are a long way from the end of fundamental advances in science, let alone being restricted to a kind of cleanup operation.

Synthesis

It is the search for theorems that puts the study of emergence squarely in the scientific domain. Although the theoretical scientist proceeds via an amalgam of discipline, modeling, selective use of observations, and sheer intuition, the end product is a rigorous (unambiguous) derivation of the consequences (theorems) of a set of inferred laws (generators). It is these derived consequences, when theory is successful, that suggest new kinds of observations. For example, Einstein's theory led at first to subtle tests, such as minor deviations in the orbit of the planet Mercury, or the slight displacement of a star's image when the sun was nearby, that could only be of interest to an aficionado. But those early tests led to much more dramatic experimental setups and observations, concerned with the derived equivalence of matter and energy and culminating in the ability to control amounts of energy orders of magnitude larger than had previously been possible.

This story has been repeated over and over again for the theories of science—not just for the major theories (gravity, electromagnetism, thermodynamics), but also for the less comprehensive theories (of superconductivity, light amplification, and so on). First come the subtle tests of little interest except to those directly involved. Then, if those implications are verified, come derivations and observations with wider implications.

Mathematical theory has been so consistently successful in this role that even scientists remark on its "unreasonable effectiveness" in helping us understand the world. Our discussions of building blocks go far, I think, toward resolving this mystery. The whole of mathematics, from numbers onward, derives from observation and modeling of the world. Indeed, it would be surprising, and it would counter everything we know about evolutionary processes, if mathematics were somehow irrelevant to the way the world works. The principal building blocks that survive and proliferate in cultural evolution (like games and numbers) must satisfy the same criteria as the building blocks of perception (the common reusable pieces into which we parse vision) or the building blocks of biochemistry (cell adhesion molecules, the Krebs cycle, and the like). All such building blocks must serve as generators that can be combined in many ways to yield viable reactions to the world.

The origin of mathematics, in number and measure, was tightly bound to practice. Ever since, mathematics has been tied to the world around us. It is only in the twentieth century that substantial numbers of mathematicians, most notably proponents of the Bourbaki school (see Davis and Hersch, 1981), have put forth the view that mathematics can and should be pursued as a completely abstract discipline, using standards completely divorced from application. Even so, most mathematicians rely on subtle intuitions tempered by earlier work in number, geometry, and physics. Mathematics embodies our most sophisticated attempts at shearing away detail to get at broadly applicable generalizations. Given this background, why should we be surprised that the building blocks provided by mathematics are generators par excellence for models of the world?

A closely related issue bears on confronting models (theories) with data: at what point should a model be directly tested against data? All models, of course, depend on observation and data. However, it can be a fatal error in the early stages of model building to try to match the data too closely. The search for mechanisms depends on a subtle selectivity, and that includes discarding as irrelevant many factors that are obvious and prominent. Aristotle's observation that physical bodies "naturally come to rest" waylaid the theory of gravity for more than a millennium. Observation and data do enter the early stages, but through wide familiarity with established observations, extant models, and interpretations, not through direct experiment.

Said another way, concentration on experimental design at the formative stage of model building brings obvious detail to the fore. That makes it difficult to apply intuition, metaphor, and all the other subtle understandings that go into insightful model building. The essence of model building is shearing away detail, but the apparatus of experimental design moves concentration in the opposite direction. Moreover, a satisfactory model, almost by definition, suggests experiments that are not at all obvious while the model is being formulated. The observation of the sun's displacement of star images requires very special conditions (an eclipse) and is not something one is likely to look for in the absence of the implications of relativity. For all these reasons the builders of models and theories should resist premature exhortations to "test it."

In making this statement, let me emphasize two caveats. First of all, at the right stage, the model *must* be tested carefully if it is to be accepted. The rigorous formulation of models, accompanied by the apparatus that uses this rigor in the design of tests, makes possible outcomes that depend less on "the eye of the beholder" than the outcomes of any other human intellectual endeavor. Herein lies a principal part of the reason for the effective, cumulative world description produced by science. The second caveat concerns the strong impulse to preview the model's implications during the early stages of model building. It is an impulse that should be acted on, but carefully. Test runs can provide valuable feedback in model (theory) synthesis, especially if one has some unusual experiments in mind. But the tests should be probes (say a run of a computer-based model) allowing program modifications along the way, or tests with easily modified, "breadboard" apparatus. Tests at this stage should *rarely* be organized attempts to collect statistics.

Emergence—Next Steps

What, then, can we expect in the study of emergence? My objective here has been to provide convincing evidence that we can increase our understanding of emergent phenomena through scientific study. Models have had center stage in this effort, and I think they will have an equally important role in the work to come.

The chief points uncovered in this exploration are consolidated in the formally defined models called *constrained generating procedures*. While developing this framework, I was continually "looking over my shoulder" to be sure that most extant models exhibiting emergence—cellular automata, Samuel's checkersplayer, neural networks, and so on—could be readily translated into the *cgp* framework. This precaution provides partial insurance against idiosyncrasy. It also incorporates knowledge already won, while assuring a modicum of generality. However, the *cgp* formalism will gain scientific standing only when it accumulates a body of theorems that act as guidelines for observing real emergent phenomena.

At this point, we are far from a theory that will enable us to derive dramatic consequences. We do, however, have a fair idea of the salient characteristics of systems that exhibit emergence, and we have an idea of the ways in which those characteristics can be abstracted to form the elements of a theory. *Cgp*'s, for instance, center on the generative powers of building blocks because most clear examples of emergence are succinctly described in such terms. We've seen repeatedly that a small set of well-chosen building blocks, when constrained by simple rules, can generate an unbounded stream of complex patterns. As Murray Gell-Mann (1994) says succinctly in *The Quark and the Jaguar:*

> Scientists, including many members of the Santa Fe Institute family, are trying hard to understand the ways in which structures arise without the imposition of special requirements from the outside. In an astonishing variety of contexts, apparently complex structures or behaviors emerge from systems characterized by simple rules.

Most human-defined systems exhibiting emergence, ranging from games to cellular automata to axiom systems, are best described in this way.

Usually the persistent patterns that arise in these generated systems are not easily anticipated on direct inspection of the generators and constraints. The most lucid examples of emergence arise when these persistent patterns obey macrolaws that do not make direct reference to the underlying generators and constraints. We've looked at the glider in Conway's automaton as a simple, fully defined example of this kind of emergence. However, the powerful uses of this technique are exemplified by scientific reductions, such as the reduction of chemical reactions to the quantum physics. There we see the full power of reduction and its obverse, description in terms of macrolaws.

For the *cgp* framework, or something similar, to acquire the status of a full-blown theory of emergence, it would have to be refined to yield sufficient conditions for emergence. We would have to prove that emergent phenomena *will* occur when these sufficient conditions are present. As it is, *cgp*'s are constructed on the premise that they exhibit enriched opportunities for emergent phenomena. My next efforts along these lines will be to provide a tighter link between the idea of *mechanism,* the basic element of *cgp*'s and the notion of *agent* (recall the ant colony). I think a minimum of three levels will be necessary to arrive at a representative body of theorems relevant to emergence: mechanism → agent → aggregate. Formally, we might consider only two levels, treating all higher levels as recursions of these basic relations. With no very sound argument to back me, I think the three-level study will be more revealing and more productive as a first step.

A useful body of theorems along these lines would link the behavior of individual agents downward to the interactions of simple mechanisms, while revealing the emergent properties of aggregates of these agents (again, recall the ant colony). It is particularly important that the theorems delineate the (new) constraints imposed by the aggregate on the individual agents. This is not at all an abstract notion: when a projected change in interest rate causes a sell-off in the stock market, we have an unambiguous example of aggregate quantities affecting the action of individual agents.

Such interactions give some hint of the departures from traditional system theory that would validate a theory of emergence. Because generated systems tend to grow in complexity over long time spans—witness the inordinate range of species produced by evolution—focus shifts from end points (optima, basins of attraction, and the like) to transitional effects (improvement, persistence, emergent constraints). The emergence of overlaid generating systems is a critical feature that does not fit into the formalism of traditional system theory. Yet, as we've seen, such an overlay can completely divert the system from its previous course. The transition from the world of individual craftsmen to the world initiated by Adam Smith's pin factory lies beyond the ken of traditional dynamics. Add to this the tremendous increase in range that occurs each time a new building block (the transistor, for instance) is discovered. A theory of emergence must offer insights into these processes and suggest conditions that enhance or suppress their occurrence.

Emergence—Horizons

To build a competent theory one needs deep insights, not only to select a productive, rigorous framework (a set of mechanisms and the constraints on their interactions), but also to conjecture about theorems that might be true (conjectures, say, about *lever points* that allow large, controlled changes in aggregate behavior through limited local action—see Holland, 1995). In the case of emergence, these insights can only be gained with the help of computer-based *exploration*. I emphasize the word "exploration" because we are not yet at the point of data matching and statistics—we need to discover *gliders*.

There are several stations along this path. First, models will serve as our chief guide in exploring prospective mechanisms and constraints. Then we can look for conditions that will assure emergence. Along the way we should learn more about predicting and controlling critical stages in the process of emergence. This knowledge, in turn, should help us to understand better the processes that crown human intellectual undertakings: innovation and creation. Let's look at each station in turn.

Models as Guides

It is usual to think of models as validated through the correctness of their predictions about the world, but we have seen at least two other roles for models in science. Models in these other roles have a different kind of validation. One role is to provide a rigorous demonstration that something is *possible*, as in von Neumann's demonstration that a machine is able to reproduce itself. Here validation occurs when a dynamic model works as claimed, as when one validates a patented device. Another role is served when the model suggests *ideas* about a complex situation, suggesting where to look for critical phenomena, points of control, and the like. Maxwell's "floating gears," Simon's 1969 limited rationality model in economics, Shrödinger's 1956 quasi-crystal model of life, and "lock and key" models in immunology are all examples of such models. They explore and explain; validation is in the cogency and relevance of the ideas they produce. Constrained generating procedures, and related models such as the Echo models (Holland, 1995), will serve first in this latter role if they are successful.

The use of computer-based models in exploration is akin to the use of gedanken experiments in physics. It is much more an intellectual exercise than a gathering of data. The rigorous definition of the computer-based models guarantees that the starting conditions are sufficient for any phenomena observed during execution. The formalism does *not* guarantee that the phenomena will recur when the starting conditions are slightly changed. However, it is easy to alter the starting conditions of a computer-based model, allowing a search for variants that *do* yield the same phenomena.

In modifying the starting conditions, and in deciding which

phenomena are critical or characteristic, insight and taste are indispensable. In the study of emergence, interdisciplinary comparisons are a critical aid in developing these two traits. As I've emphasized in several places, interdisciplinary comparisons allow us to differentiate the incidental from the essential. When we look for the same phenomena in different contexts, we can separate features that are always present from features that are tied to context. Moreover, what is hidden in one context may be obvious in another. Both aspects offer great help in the construction of models of emergence.

The search at this stage is a matter of intelligent probing, not a matter of runs and reruns yielding "statistically significant relations." The Baconian approach of gathering data until significant relations emerge is unlikely to work because systems exhibiting emergence are so complex. In particular, nonlinear interactions almost always forestall the simple parsings, such as regressions, that can reveal causal relations. Careful probing, moving back and forth between conditions and phenomena, offers a way of obtaining the outlines of significant relations. Such relations, suggesting conditions and their consequences, provide the grist for a more general, formal theory.

The exploratory phase should lead in time to an understanding of conditions that *assure* emergence. If, at this early stage in our studies, I were to postulate a set of conditions sufficient for emergence, those conditions would center on the conditions that give rise to complex adaptive systems. Here there is a reciprocal relation: we will not truly understand complex adaptive systems until we understand the emergent phenomena that attend them. In our exploration of complex adaptive systems, computer-based models have already played a critical role. They have shown that agents obeying simple rules can mediate sophisticated trade/combat interactions. The caterpillar-ant-fly triangle, the evolution of cooperation in the Prisoner's Dilemma, and markets based on agents using simple buy/sell rules have substantially broadened our understanding of the kinds of behavior typical of complex adaptive systems. This growing understanding will, inevitably, help us formulate sufficient conditions for emergence. Eventually we should be able to surround the rigorous definition of such structures with a body of theorems that serve as guidelines for experiment, much as Maxwell's equations and theorems taught us to look for electromagnetic waves and spectra.

At the broadest level, this book's thesis has been that models and model building underlie much, perhaps most, human intellectual endeavor. At every point along the way, we have encountered models, ranging from early games to rigorous mathematical models and, more recently, computer-based models. We have discussed the relationship between target and source models as a way of understanding the kinds of reduction that are so productive for science. Above all, we've seen that models give us a way of compensating for the perpetual novelty of the world.

Control and Prediction

By inferring lawlike connections between salient, repeating features of the world, we can bring past observations to bear on current conditions. We can even anticipate and control future occurrences.

Directors of large corporations and government bodies now routinely explore options by using models, such as spreadsheets, based on linear trend analyses. Some models go beyond this, employing IF/THEN rules and allowing a range of control options, much like a flight simulator. It is *not* common, but it is important, to allow those same directors to see and manipulate the mechanisms and interactions *underlying* such models, using their intuition to move the models into plausible regimes. When this is possible, the model often reveals certain disastrous "cliffs" that appear and reappear under a wide variety of assumptions. In such situations the model does indeed act like a flight simulator, letting us "push the envelope" without committing us to overt actions that will lead to disastrous consequences. In helping us avoid these cliffs—disastrous, irreversible situations—it is almost always the case that a well-designed model is better than no model at all. The model can shed some light on the situation, and large cliffs are likely to loom even in a dim light. Murray Gell-Mann makes this point nicely in *The Quark and the Jaguar* (1994).

> [A] multiplicity of crude but integrative policy studies, involving not just linear projection but evolution and highly nonlinear simulation and gaming, may provide some modest help in generating a collective foresight function for the human race. . . . We are all in a situation that resembles driving a fast vehicle at night over unknown terrain that is rough, full of gullies, with precipices not far off. Some kind of headlight, even a feeble and flickering one, may help to avoid some of the worst disasters.

As Berliner (1978) points out with respect to games, if we can avoid the cliffs we retain the opportunity to correct less disastrous situations.

Innovation and Creation

The close relation between emergence and innovation comes to the fore when we look at "the supreme Figure of all: Metaphor" (Eco, 1994). The source/target construction of metaphor and the source/target form of reduction in science are closely related, and both have a central role in viewing the world anew. We have seen that creation in both science and literature depends on a disciplined sensitivity to the aura of technique and meaning that surround source and target, be they models or objects. Ultimately, significant innovation depends on the "long line": the ability to go beyond cut-and-try recombinations of well-known building blocks to the more distant combinatorial horizon.

In both the arts and the sciences, the directions of innovation are set by constraints and bottlenecks. The bottlenecks imposed by technical difficulties (constraints) make some combinations difficult or impossible. Under these circumstances the horizon becomes the next hill, not a distant vista. It is here that the distinction between innovation and optimization looms large. Optimization in complex adaptive systems is rarely possible, and it is often not even meaningful. What would be the optimal organization for an animal inhabiting a tropical forest? Significant innovation requires discovering a combination that is intermediate between obvious cut-and-try and the infeasible optimum. The amalgam of interdisciplinary comparison, immersion in relevant discipline, and sensitivity to relevant auras serves us well at this stage. In contrast, "turning the deductive crank" has as little relevance in this part of the scientific endeavor as it does in the arts.

Though it is thought of as commonplace in the arts, "getting more out than you put in" goes against intuition in the sciences. Nevertheless, there is a real sense in which this occurs in systems that exhibit emergence. We've seen that systems with compact, lawlike definitions, from games to scientific theories, can be studied for extended periods without exhausting their content. Of course, the possibilities for the system are fully determined once we set down the definition. When the system involves IF/THEN conditionals or other nonlinear interactions, though, direct inspection of the definition does not reveal these possibilities.

In contemplating this aspect of emergence, it is important to concentrate on the full-fledged version of reduction that takes interactions into account. "Reduce the system to parts and, once you understand the parts, you will understand the system" is simpler and often works in the sciences—science has advanced because it does work. But this simpler version does not work for systems exhibiting emergence. Instead, the more difficult form of reduction, using overlaid hierarchies (the discovery of *gliders*), is the productive way to approach such systems. This more difficult form of reduction is also typical of the broader reaches of science. The laws of chemistry are indeed constrained by the laws of physics and, in this sense, chemistry is reducible to physics. However, chemistry has its own *gliders*—molecules. The macrolaws that govern the interaction of molecules are formulated and used without reference to the laws of particle physics. In unusual circumstances chemists refer to deeper levels (such as the effects of radioactivity), but these are the exception not the rule.

Emergence, then, is a matter of macrolaws and the overlaid constrained generating procedures that result. A detailed knowledge of the repertoire of an individual ant does not prepare us for the remarkable flexibility of the ant colony. The capabilities of a computer program are hardly revealed by a detailed analysis of the small set of instructions used to compose the program. We will soon know the complete set of genes (or at least some of the alleles of each gene) coded in human DNA, but we will be far from understanding the program those genes specify—the program that takes a fertilized egg to the complicated 100-billion-cell mature organism. It is difficult to understand the interactions and emergent phenomena, the stuff of biology and medicine, that attend this vast ecosystem. But there is more. In that array of more than 100 billion cells is a network consisting of several tens of billions of specialized cells called neurons. Understanding the behaviors mediated by these cells, the stuff of psychology, is much more than a matter of understanding the properties of isolated neurons.

The fact that systems exhibiting emergence require studies that go beyond the simple reduction of studying isolated parts does *not* mean that those systems are beyond our grasp. After all, chemistry is a *very* successful science, even though we cannot understand that science via a direct investigation of the laws of physics. Patience is required. Games like chess and Go, with defining rules so simple they are quickly comprehended by a young child, have been studied for centuries, and still we learn. Why should we expect it to be different with the more intricate rules that define complex adaptive systems and other systems that exhibit emergence?

Emergence—Far Horizons

A deeper understanding of emergence will help us with two profound scientific problems that have philosophic and religious overtones: *life* and *consciousness*. We've already looked at the resolution of a similar, though simpler, problem: can a machine reproduce itself? Life and consciousness are both more difficult and more far-reaching in their implications.

Consider *life*, that grand abstraction we use to separate the entities of this world into two distinct classes: the living and the nonliving. There is a bit of paradox in this classification. We have no reservations about calling molecules "dead," but the aggregate of such molecules, organized as a biological cell, is certainly "living." Most scientists now believe that there is no hidden "vital" principle that somehow supersedes the laws of physics and chemistry. Still, we have no models that rigorously exhibit life as an emergent phenomenon, only conjectures about what such models would look like (Eigen and Winkler, 1981). On good evidence, we believe that chromosomes set the master plan for the development of an organism, and we begin to know the functions of the individual genes. But we know remarkably little about the ways in which the genes interact. It is evident that the chromosome specifies a complicated program. Genes are turned "off" and "on" during the development of a cell, on the basis of complicated feedbacks mediated by cell proteins. For a metazoan cell, this program is almost certainly more complicated than any program we have written for a programmed computer. Moreover, this program is immersed in a complicated milieu of structures and catalysts, which it modulates and which modulate it. While almost all scientists believe that life is an emergent phenomenon, there is a long way to go before we have working models based on known mechanisms.

Though most scientists believe life is an emergent phenomenon, the same cannot be said for *consciousness*. Despite sustained deliberation almost from the beginning of recorded history, the nature of consciousness is still an open question. It is unclear whether any, or all, of consciousness can be reduced to the interactions of neurons. Interesting conjectures exist (see Dennett, 1991, for a particularly interesting discussion), but no theory or model that rigorously exhibits consciousness as an emergent phenomenon. Indeed, we have neither theories, models, nor artifacts wherein each agent (neuron) *simultaneously* interacts with thousands of other agents (via synapses), and wherein the connections among agents involve so many feedback loops that a single agent may belong to hundreds or thousands of loops. Even an intricate computer provides only ten or so contacts for each component. To extrapolate from our current knowledge of machines to such machines is to jump three orders of magnitude in complexity. What we now know of machines provides little guidance to machines of that complexity. Extrapolation in such circumstances is, at best, speculation. Until we know more about machines of this complexity, whether or not consciousness can be understood as an emergent property of certain kinds of machines is moot.

Our understanding of the universe will be severely limited until we have a more definitive view of how much of *life* and *consciousness* can be explained as emergent phenomena. We must know how far we can go with explanations based on the interactions of a few well-understood mechanisms (say biomolecules and neurons, respectively). We are a long way from knowing the limitations (if any) of such explanations. But until we have made a sustained effort at such an explanation, we will not know what must be explained in other ways.

At a still further remove is the guiding question of this whole work: how can the interactions of agents produce an aggregate entity that is more flexible and adaptive than its component agents? It is not an impossible question, and answers to it are certainly subject to scientific tests. It is a difficult question, and one that will require sustained effort over a long period. Whatever answers we come upon will profoundly affect our view of ourselves and our world.

References

(* Indicates reference accessible to the general reader.)

Arthur, B. W., J. H. Holland, et al. 1997. "Asset Pricing under Endogenous Expectations in an Artificial Stock Market." Santa Fe Institute Working Paper 96-12-093.

Axelrod, R., and W. D. Hamilton. 1982. "The Evolution of Cooperation." In J. Maynard Smith, ed., *Evolution Now*. New York: Freeman.

*Berliner, H. J. 1978. "A Chronology of Computer Chess and Its Literature." *Artificial Intelligence* 10(2).

*Black, M. 1962. *Models and Metaphors*. Ithaca, N.Y.: Cornell University Press.

*Borges, J. L. 1970. *The Aleph and Other Stories*. New York: Dutton.

Burington, R. S. 1946. *Handbook of Mathematical Tables and Formulas*, 2nd ed. Sandusky, Ohio: Handbook Publishers.

*Burke, J. 1978. *Connections*. Boston: Little, Brown.

Chomsky, N. 1957. *Syntactic Structures*. The Hague: Mouton.

Cowan, G. A., D. Pines, and D. Meltzer. 1994. *Complexity: Metaphors, Models, and Reality*. Reading, Mass.: Addison-Wesley.

*Davis, P. J., and R. Hersh. 1981. *The Mathematical Experience*. Boston: Houghton Mifflin.

*Dennett, D. C. 1991. *Consciousness Explained*. Boston: Little, Brown.

*———. 1995. *Darwin's Dangerous Idea*. New York: Simon & Schuster.

*Eco, U. 1994. *The Island of the Day Before*. New York: Harcourt Brace.

Eigen, M. and R. Winkler. 1981. *Laws of the Game*. New York: Knopf.

Feynman, R. P., et al. 1964. *Lectures on Physics*. Reading, Mass.: Addison-Wesley.

*Gardner, M. 1983. *Wheels, Life, and Other Mathematical Amusements*. New York: Freeman.

*Gell-Mann, M. 1994. *The Quark and the Jaguar: Adventures in the Simple and the Complex*. New York: Freeman.

Hebb, D. O. 1949. *The Organization of Behavior: A Neuropsychological Theory*. New York: Wiley.

*Hesse, Herman. 1943. *Das Glasperlenspiel.* Translated (1969) as *Magister Ludi.* New York: Holt, Rinehart and Winston.

*Hesse, M. B. 1966. *Models and Analogies in Science.* South Bend, Ind.: Notre Dame University Press.

*Hofstadter, D. R. 1979. *Gödel, Escher, Bach: An Eternal Golden Braid.* New York: Basic Books.

*————. 1995. *Fluid Concepts and Creative Analogies.* New York: Basic Books.

*Holland, J. H. 1995. *Hidden Order: How Adaptation Builds Complexity.* Reading, Mass.: Addison-Wesley.

Jammer, M. 1974. *The Philosophy of Quantum Mechanics.* New York: Wiley.

Kleene, S. C. 1951. "Representation of Events in Nerve Nets and Finite Automata." In C. E. Shannon and J. McCarthy, eds., 1956, *Automata Studies.* Princeton: Princeton University Press.

Korth, J. J., ed. 1965. *IBM Scientific Computing Symposium: Large-Scale Problems in Physics.* White Plains, N. Y.: IBM.

May, R. M. 1973. *Stability and Complexity in Model Ecosystems.* Princeton: Princeton University Press.

Maynard Smith, J. 1978. *The Evolution of Sex.* Cambridge: Cambridge University Press.

Maxwell, J. C. 1890. *The Scientific Papers of James Clerk Maxwell.* Cambridge: Cambridge University Press.

Minsky, M., and S. Papert. 1988. "Perceptrons." In J. A. Anderson and E. Rosenfeld, eds., *Neurocomputing.* Cambridge, Mass.: MIT Press.

Misner, C. W., K. S. Thorne, and J. A. Wheeler. 1970. *Gravitation.* San Francisco: Freeman.

Mitchell, M. 1993. *Analogy-Making as Perception.* Cambridge, Mass.: MIT Press.

————. 1996. *An Introduction to Genetic Algorithms.* Cambridge, Mass.: MIT Press.

*Newman, J. R. 1956. *The World of Mathematics.* New York: Simon & Schuster.

Rashevsky, N. 1948. *Mathematical Biophysics.* Chicago: University of Chicago Press.

Rochester, N., J. H. Holland, et al. 1956. "Tests on a Cell Assembly Theory of the Action of the Brain, Using a Large Digital Computer." In J. A. Anderson and E. Rosenfeld, eds., 1988, *Neurocomputing.* Cambridge, Mass.: MIT Press.

Samuel, A. L. 1959. "Some Studies in Machine Learning Using the Game of Checkers." In E. A. Feigenbaum and J. Feldman, eds., 1963, *Computers and Thought*. New York: McGraw-Hill.

*Schrödinger, E. 1956. *What Is Life?* New York: Doubleday.

*Sholl, D. A. 1956. *The Organization of the Cerebral Cortex*. London: Methuen.

*Simon, H. A. 1969. *The Sciences of the Artificial*. Cambridge, Mass.: MIT Press.

*Singer, C. 1959. *A Short History of Scientific Ideas*. Oxford: Oxford University Press.

Turing, A. M. 1937. "On Computable Numbers, with an Application to the Entscheidungsproblem." *Proceedings of the London Mathematical Society*, series 2, no. 4:230–265.

Ulam, S. 1974. *Sets, Numbers, and Universes*. Cambridge, Mass.: MIT Press.

von Neumann, J. 1966. *Theory of Self-Reproducing Automata*, ed. A. W. Burks. Urbana: University of Illinois Press.

von Neumann, J., and O. Morgenstern. 1947. *Theory of Games and Economic Behavior*. Princeton: Princeton University Press.

Index